Meteorology and Physiology in Early Modern Culture

Meteorology and Physiology in Early Modern Culture: Earthquakes, Human Identity, and Textual Representation provides the first sustained examination of the foundational set of early modern beliefs linking meteorology and physiology. This was a relationship so intimate and, to us, poetic that we have spent centuries assuming early moderns were using figurative language when they represented the matter and motions of their bodies in meteorological terms and weather events in physiological ones. Early moderns believed they inhabited a geocentric universe in which the matter and motions constituting all sublunary things were the same and that therefore all things were compositionally and interactively related. What physically generated anger, erotic desire, and plague also generated thunder, the earthquake, and the comet. As a result, the interpretation of meteorological events, such as the 1580 earthquake in the Dover Strait, was consequential. With its radical and seemingly spontaneous shaking, an earthquake could expose inconvenient truths about the cause of matter and motion and about what, if anything, distinguishes humans from every other thing and from events. *Meteorology and Physiology in Early Modern Culture* reveals a need for reexamination of all representations of meteorology and physiology in the period. This reexamination begins here with a focus on the Titanic metamorphoses captured by Edmund Spenser, William Shakespeare, John Donne, and the many writers responding to the 1580 earthquake.

Rebecca Totaro is a Professor of English at Florida Gulf Coast University.

Perspectives on the Non-Human in Literature and Culture

In recent years, many disciplines within the humanities have become increasingly concerned with non-human actors and entities. The environment, animals, machines, objects, weather, and other non-human beings and things have taken center stage to challenge assumptions about what we have traditionally called "the human." Informed by theoretical approaches like posthumanism, the new materialisms (including Actor Network Theory, Object-Oriented Ontology, and similar approaches), ecocriticism, and critical animal studies, such scholarship has until now had no separate and identifiable collective home at an academic press. This series will provide that home, publishing work that shares a concern with the non-human in literary and cultural studies. The series invites single-authored books and essay collections that focus primarily on literary texts, but from an interdisciplinary, theoretically-informed perspective; it will include work that crosses geographical and period boundaries. Titles are characterized by dynamic interventions into established subjects and innovative studies on emerging topics.

Birds and Other Creatures in Renaissance Literature
Shakespeare, Descartes, and Animal Studies
Rebecca Ann Bach

Race Matters, Animal Matters
Fugitive Humanism in African America, 1838–1934
Lindgren Johnson

Plants in Contemporary Poetry
Ecocriticism and the Botanical Imagination
John Charles Ryan

Meteorology and Physiology in Early Modern Culture
Earthquakes, Human Identity, and Textual Representation
Rebecca Totaro

Meteorology and Physiology in Early Modern Culture

Earthquakes, Human Identity, and Textual Representation

Rebecca Totaro

Routledge
Taylor & Francis Group
New York London

First published 2018 by Routledge
52 Vanderbilt Avenue, New York, NY 10017
2 Park Square, Milton Park, Abingdon, Oxon OX14 4RN

First issued in paperback 2020

Routledge is an imprint of the Taylor & Francis Group, an informa business

Library of Congress Cataloging-in-Publication Data
A catalog record for this book has been requested

ISBN 13: 978-0-367-66736-8 (pbk)
ISBN 13: 978-1-138-09216-7 (hbk)

Typeset in Sabon
by Apex CoVantage, LLC

For we three monsters, Vic, Mike, and Beck

Contents

Acknowledgments

Thanks first to 2006–2007 Folger Institute Year-Long Colloquium members and director who helped me shape the first stages of this project in the year prior to and including my sabbatical from Florida Gulf Coast University. Additional review of versions of each chapter occurred at or in communications following content-related national and international conferences, including meetings of the Shakespeare Association of America, Renaissance Society of America, Sixteenth Century Society, World Shakespeare Congress, Modern Language Association, Spenser Society and John Donne Society. Florida Gulf Coast University staff, faculty, administration, and students also read drafts; contributed to content by way of scholarly and in-class discussion of related topics; and/or offered support for conference travel. Debts of gratitude are for special colleagues who engaged with me over months of scholarly writing exchange: Delphine Gras, Kimberly Jackson, Mary Crone-Romanovski, Masami Sugimori, Fiona Tolhurst, Tim Sutton, and Laci Mattison. And to graduate student readers Susan Rojas, Kelsey Fischell, Savannah Jensen, Morgan Souza, and undergraduates Clare Brinkman, Kelsey Luft, and Chandler Rae Tarquino. Support across the university also has come generously from librarian faculty Rebecca Donlan, Rachel Cooke, and Rachel Tate-Ripperden, and from deans Bob Gregerson, Aswani Volety, and Donna Henry. My warmest appreciation for targeted, generous content suggestions after a full reading of the manuscript is for Gail Kern Paster, Katherine Eggert, Susan Staub, and Mary Thomas Crane. Readers of content over time in conference paper and seminar form have also helped me improve the whole; special thanks here to Mary Trull, Mary Floyd-Wilson, Darryl Chalk, Amy Tigner, Karen Raber, Jennifer Munroe, Rebecca Laroche, Garrett Sullivan, Margaret Healy, Shanyn Altman, Tiffany Jo Werth, Lyman Tower Sargeant, and Richelle Munkhoff, as well as to the several anonymous external reviewers selected by Routledge and, again, to Karen Raber, editor of the Routledge Series *Perspectives on the Non-Human in Literature and Culture*, and editorial staff Jennifer Abbott, Alison Daltroy, Veronica Haggar, Autumn Spalding, Megan Hiatt, and (until recently with Routledge) Liz Levine. And with much love from Keith, Vicky and Michael, Mom and Dad, Marta, Sylvia, Chris and Starr, Chris and Phil, and (meow meow) Liam and Lucas. It takes a village (and thank goodness for that).

Permissions

Early portions of chapter 3 appear in print as "The Meteorophysiology of the Curse in Shakespeare's first tetralogy," *English Language Notes.* 51.1 (2013): 191–210; and as "'Revolving This Will Teach Thee How to Curse': Lessons in Sublunary Exhalation," in Jennifer C. Vaught, editor, *Rhetorics of Bodily Disease and Health in Medieval and Early Modern England* (New York: Routledge, 2010), 135–151. An early portion of chapter 3 appears as "Britomart's Meteorological Wound" in *Archiv für das Studium der Neueren Sprachen und Literaturen* 250 (2013): 42–65. Permission has been attained for publication of whatever portion of these materials is here appearing in identical form to those already published essay versions.

A Note on Original Texts

Primary source spelling for all early modern texts has been retained, except in the following cases: regularization for letters, spellings, and punctuation marks that would render comprehension too difficult if left uncorrected, such as the choice to transpose for u/v, i/j, and s/f; omission of accent marks; when comprehension and/or clarity necessitated modernization; or when I am citing a work already modernized in the source consulted, as for the plays of Shakespeare. The other exception is in the case that a current translation retains original spellings including use of letters I would otherwise transpose, such as for Hamilton's *Faerie Queene.*

Introduction

> [T]here was an era when the body represented something quite different from the entity we imagine now—a discrete given, an independent and isolated object. Once upon a time, all reflection on what we call the body was inseparable from inquiry into places and directions, seasons and winds. Once upon a time, human being was being embedded in a world.
>
> Shigehisa Kuriyama, *The Expressiveness of the Body and the Divergence of Greek and Chinese Medicine*[1]

Examining the human experience of embeddedness in a physical world constituted by earth, wind, fire, and water in motion, I predicate this book on a foundational set of early modern beliefs about the relationship between meteorology and physiology. This is a relationship so intimate and, to us, poetic, that we have spent centuries assuming that early moderns were speaking figuratively when they represented the matter and motions of their bodies in meteorological terms and weather events in physiological terms. Early moderns believed that they inhabited a geocentric universe in which the matter and motions constituting all sublunary bodies and things were essentially the same and that therefore the forms taken by sublunary matter were literally, interactively, and micro-macrocosmically related. What made and fueled anger, orgasm, and plague also made and fueled thunder, the earthquake, and the comet. Correspondingly, categories of human and animal and of biotic and abiotic, as well as of science and poetry, politics, and astronomy were not as distinctly demarcated as they are now. In other words that put a fine point on Kuriyama's informed observation opening this introduction, "[o]nce upon a time, human being" mixed interactively with all sublunary being and perceived through its senses that this mixing, was natural, even as it sought through philosophy and religion to ponder the degree to which humans must also be unique. *Meteorology and Physiology in Early Modern England* examines for the first time at length the some of most essential indicators of this experience of human being, made manifest in the textual representation of the relationship between meteorology and physiology.

As events more than as things, bodies in Kuriyama's "[o]nce upon a time" resembled the weather in their openness, with outside and inside in continual exchange. Drawing from research by Mary Floyd-Wilson and by Gail Kern Paster, Tanya Pollard, and Katharine Craik explain further: "The humoral body was implicated in a network of sympathies with the wider world where the cosmic macrocosm was understood to be as sensible—and as vulnerable to change—as the subjects who lived in it."[2] In textual representation, early moderns wrote of feeling and of understanding the psychophysiological matter and motions of their bodies in the very ways they felt and understood the matter and motions in their environments, and vice versa.[3] For these reasons, the term I have coined—"meteorophysiological"—is useful for describing this early modern condition of embeddedness in multiple, interactive, complex systems. The early modern condition of being, I argue, is a meteorophysiological one as much as it is a religious or a pre-scientific one. Viewing the early modern sublunary system and its subjects as thus meteorophysiologically constituted and interactive gives us new reason to return with fresh eyes to the works of all writers from the time who, with their readers, belonged to a world in which the same unwilled, complex forces fueling the earth to produce the earthquake, tidal wave, volcano, comet, and thunderclap also fueled their bodies in the production of its symptoms of change, from its anger and laughter to its plague buboes and love-wounds.[4] In the imaginative literature of Edmund Spenser, William Shakespeare, and John Donne, the seemingly infinite representational potential of this relationship between meteorology and physiology gained its most resonant and enduring expression, urging us to venture beyond the confines of our own paradigms where there may be monsters but where there will always also be the promise of change and the delights of play.

*

In *Tamburlaine the Great*, first performed in 1587, Christopher Marlowe gives us an apt picture of microcosm and macrocosm, meteorology and physiology in just such intense, mutually transformative interplay. Here Tamburlaine predicts the change that will come upon a set of countrymen when they join up with his full army:

> . . . these that seem but silly country swains
> May have the leading of so great an host
> As with their weight shall make the mountains quake,
> Even as when windy exhalations,
> Fighting for passage, tilt within the earth.[5]

In this passage, the meteorological potential of the army to "make the mountains quake" grants these men, collectively, the power of the Old

Testament God, of Poseidon the earth-shaker, of the Titans, and of Britain's mythological giants. Marlowe enhances the description by way of a simile, adding more air, "as when windy exhalations/Fighting for passage, tilt within the earth." This appears to be a straightforward simile that is lovely and fitting. Upon closer inspection, however, it might seem odd to shift registers from power vested in great weight, as of a Titan, to power vested in something that to us is comparatively slight: "windy exhalations." Within the context of early modern meteorology and physiology, however, these exhalations are among the most essential and active of materials and motions in the sublunary system. They are also substances that, as I will explain in chapter 1, appear in Aristotle's *Meteorology.* What the exhalations do here is act just as the human passions do. They inform the bodily interiors of the "country swains" and unite them in zeal by newly "tilt[ing] within," bringing each into altered states of being and into a condition of resemblance and shared motion enough to "make the mountains quake." This is Shakespearean "sea-change"[6] across many men, Ovidian metamorphosis multiplied, as the soldiers come to see themselves not only with new eyes but with all senses and all relationships—to themselves, to each other, to each subsequently encountered thing in the world including the elements—transformed. We are not encountering "the weather" as we know it now. Early modern meteorology was a different creature altogether, informed by geohumoral theory and centuries of literary tropes placing meteorology and physiology together in responsive, fully integrative interactions. The result was a sensual experience of the human body and of the full sublunary sphere that is difficult to access but available through the period's visual and textual representations. In England at this time, the dominant expressive medium was text, versus the image, and for this reason here textual representation emerges as the primary reservoir of evidence.

Shakespeare's description of the transformative power of zeal upon individuals and groups in his history play *Henry V* (1599) also gives voice to the full sensory and ecosystemic involvement of the meteorophysiological experience. In directions given at Harfleur to his men, who are outnumbered and in fear for their lives, Shakespeare's Henry V instructs them in the ways of alteration that are at once meteorological and physiological, human and non-human:

> [W]hen the blast of war blows in our ears,
> Then imitate the action of the tiger;
> Stiffen the sinews, conjure up the blood,
> Disguise fair nature with hard-favour'd rage;
> Then lend the eye a terrible aspect;
> Let it pry through the portage of the head
> Like the brass cannon; let the brow o'erwhelm it
> As fearfully as doth a galled rock

O'erhang and jutty his confounded base,
Swill'd with the wild and wasteful ocean.
Now set the teeth and stretch the nostril wide,
Hold hard the breath and bend up every spirit
To his full height.[7]

Shakespeare depicts the pressure of air held within a body as fuel for alteration. Mountains are not moved but the body will show itself as a "tiger," transformed through its breathing pent—"h[e]ld hard"—into something it was not. Henry's men and Tamburlaine's are set to become not only physical armies of might but also meteorological ones, harnessing the literal forces that move the elements of the sublunary system. Both leaders teach their men how to perform themselves into difference, into a new disposition, and even literally into a new composition of air and other elements. This feat constitutes more than acting as we would know it. These leaders direct their men through a process of material alteration and becoming that is both psychophysiological, to use Paster's term, and meteorophysiological, wherein the psychological is assumed.

In applied terms, the zeal of these men is literally a matter of the humors and the elements.[8] Because of this, Shakespeare and Marlowe's military men are also both agents of and subjects of change in this scenario, as they interact with their full sublunary environment. This full interaction is in evidence through the diversity of associated components the playwrights use to describe the transformation of these regular men into soldiers. To bring out such change involves human bodies, weight, mountains, the blast of war, sinews, blood, rage, eyes, teeth, nostrils, breath, and spirit. These things appear as having a literal correspondence with the action of the tiger, fighting exhalations, brass cannon blasts, the ocean, and earth's body shaking. The whole set, greater than the sum of its parts, is in each text the physiological equivalent of an earthquake experience that alters each member, the set itself, and all with which they will come to interact. These armies will deplete resources, for example, thus reconfiguring the exchanges among people whose lands they cross or conquer, and, as their leaders hope, these armies will redraw political and perhaps even geological boundaries, as an earthquake could.

What we witness above has relationship to the current concept of the "assemblage," coined by Gilles Deleuze and Felix Guattari and, here, described by new materialist and political scientist Jane Bennett as a "human-nonhuman working group" in which all things in the group are "actants" and the group together exerts a distributive rather than localized agency.[9] Such an assemblage is "as much wind as thing, impetus as entity, always on the way to becoming otherwise, an effluence that is vital and engaged in trajectories but not necessarily intentions" (119). The transformative nature of the assemblage is what is at stake in Marlowe's and Shakespeare's representations of the metamorphoses of their men. The men undergo more than a physiological or even psychophysiological alteration. They experience a change involving the physiological and meteorological, biotic and abiotic, human and non-human, part and whole, individual and army, internal and external. None of this change is uniform, but rather it is unique for each thing in the constantly altering system of the assemblage. None of this change, further, is willed. It is more a result of natural trajectories than of intentions, more about principles related to all things material and all events derived from them. In Jane Bennett's words, a world of vital matter seen active by way of assemblages is "a world filled not with ontologically distant categories of being (subjects and objects) but with variously composed materialities that form confederations" (99). In this way, Bennett's attempts to imagine how it feels to experience vital materialism suits the early modern experience of a meteorophysiological body.

*

Examining this body as it anticipates and undergoes the radical change of an earthquake specifically highlights for closer inspection this experience of embodiment. In John Webster's *The Duchess of Malfi*, for example, a swift terrifying realization of doom calls upon Antonio and the Duchess as they greet multiple, unanticipated, visitors, each of whom delivers a new form of terror:

Antonio [Knocking within.]
How now! who knocks? More earthquakes?
Duchess. I stand
As if a mine beneath my feet were ready
To be blown up

Antonio does not mean literally "more earthquakes" are "[k]nocking" at the door, but he does mean that he and the Duchess are in peril of experiencing irremediable psychophysiological upheaval. The Duchess then responds using a simile to express her terror, the sensation of being not generally under assault but of being psychophysiologically in peril, of waiting for the next event that might render the present condition unrecognizable. Her reference to "a mine beneath my feet" is itself an interesting extension of Antonio's reference to earthquakes. It draws on military terminology for the word "mine." The mine is not a landmine as we know it but a cavern filled in wartime with explosives that will go off without the enemy being aware.[10] Webster also gives her in this response evidence of the knowledge of Aristotelian earthquake theory, based on the idea that earthquakes occur when air is trapped in the Earth's hollows and erupts out. Our attention to the exchange is drawn by employment of pathos created from a shared understanding of the fear of falling, of suddenly losing one's footing. We imagine with these characters the terrifying threat of their impending sufferings and death in this tragedy that can have no other outcome. The terror, even in tragedy, is in the anxious fear of what cannot in its details be anticipated, of change so dynamic and thoroughgoing that it is an event not a thing, systemic not isolated.

Just so, earthquakes cause all bodies to tremble unexpectedly with promise only of alteration that cannot be entirely predicted. What is certain is that for the time, earthquakes seemed to come about due to changes in matter and motions ultimately shared with the very bodies they shook. Additionally disconcerting, beliefs about what caused and halted bodies in their trembling toward rupture were changing as well. What was it exactly that caused anger? What exactly caused earthquakes? Humanism, advances in world exploration and technology, and the Protestant Reformation opened unexplained phenomena for new consideration. In these many ways, earthquakes exemplify the complex relationship between meteorology and physiology. In the early modern period, or the premodern more generally, an earthquake could not ever be a number on a Richter scale or a factor of continental plate shifts, as it is for us. A premodern earthquake involved broad interspecies, human/non-human, macrocosmic-microcosmic interaction that was at once wondrous and known by the body's most basic senses. To account for it required narratives that allowed for its power and range, some of them drawn from the realm of myth and legend, such as those associating earthquakes with memory and with the earth-born Titans of Greek polytheism who contested the authority of the gods and were always on the verge of breaking out of their chthonic prison. Full-body memories of the quake—and what other kind can there be for those who experience them?—are just such Titans. For these reasons, in the premodern West, and in early modern England in particular, the wonder of an earthquake was readily associated with the hazards and promises of resistance, with inconvenient irreducible truths about material change writ large, and with the blurring of boundaries otherwise separating human and non-human, the living and the nonliving, physiology and meteorology, past and present.

In any era, an earthquake is an event that threatens full-body and exosystemic change and can leave a literal impression in the body and the mind, creating complex, visceral memories. In England, where earthquakes are rare, a 1580 quake shook parts of England and still more forcefully challenged the status quo. In other words, it called into question what it was to be stable as a nation, a Christian, and even a human being. In its wake arose questions about the causes and purposes of creation itself. What was preventing earth from coming apart, and if earth could shake so suddenly and then cease without any warning or reason, how and why did anything cohere? How could one be assured in the promise of a resurrected body and its union with the soul when the terror of a quake could move one to a state of doubt? The shaking of an earthquake exposed weaknesses in architectural, religious, and political structures alike. In the case of the latter exposures, the interpretation of the severity of damage often was a matter of one's faith and specific confession: Catholic versus Protestant, Church of England versus Puritan. An interpretation of the 1580 earthquake could show one to be in favor of or against the marriage of Queen Elizabeth I to the Duke of Anjou who was courting her at the time from France, for example. It might reveal whether one believed England and the faithful of England were or were not on the right side of God.

Adding to the questions an earthquake might elicit in those who experience it, and considering the complexly related categories of knowing and ways of being an earthquake required in answer to those questions, a turn to the Bible for answers could either settle all matters or exacerbate them, because biblical accounts of earthquakes are one-note wonders that do not suit experience. Biblical accounts of earthquakes render them as interchangeable with all radical sublunary alterations, like plague and famine, one of so many just punishments issued by God upon a forgetful people in need of constant warnings. In other words, an earthquake may impact a nation such that the quake appears as one of God's many messages to all inhabitants, but those inhabitants will have various immediate and long-term feelings and thoughts related to a quake, and each feeling and thought will be the product of individual life contexts irreducible to the static biblical accounts. Each feeling and thought will additionally change over time when replayed through memory. The experience of an earthquake was at once personal and national; externally and internally altering; and physiological, meteorological, social, religious, and political in impact. It thus caused effects striking in resemblance to those experienced by the humans and animals represented in Ovid's *Metamorphoses* who undergo radical alteration that begins in trembling.

Beyond Ovid and the Protestant propaganda pamphlets of monstrous births, however, a trembling earth could resonate with many bodies in as many ways. It might be associated with the happy, loving memories, such as those of Juliet's nurse at her weening or of Glendower's mother at his birth, subjects of chapter 5; it might instead relate emasculation in the disintegration of the once-potent body, such as that often captured by John Donne in *Devotions Upon Emergent Occasions*, the subject of chapter 6. Such is the mystifying power of the earthquake, of the impression it makes uniquely on all sublunary, meteorophysiological bodies. It cannot be that one account of its cause or one experience of its effects entirely suffices to render the event moderately comfortable, let alone comprehensible. In eruptive trembling of this kind are the individually generative seeds for stories of metamorphosis, of the body rumbling toward change through fear, zeal, and anger but also through erotic love, maternal power, and genial laughter. These catalytic emotions and actions can transform bodies, even from one state into another, but never in the same way twice.

To be the writer representing an earthquake in early modern England was, then, to be the writer asserting claims related to a much wider range of subjects than would now be generally accepted into its consideration. To write about the earthquake was to proclaim one's beliefs regarding on God's relationship to humans and to his created world in general. It was to show oneself leaning Catholic or Puritan, open to some of the teachings of Aristotle regarding the cause and meaning of *meteora* (a term I use here to name all meteorological phenomena identified by Aristotle). Aristotle's teachings had troubled the church for centuries and to proclaim on the earthquake was therefore also to take a public position on the stability of a sovereign's reign and thus on national security. It was to capture in words the most pervasive of metamorphic experiences. To represent the earthquake was to comment on the nature of change itself, the promise of the rainbow, and details of the Apocalypse, its very how and why and even exactly when it shall come. For these reasons, to write about the earthquake was to enter into thrilling territory with enormous expressive potential and with risk. As Thomas Dekker writes in *The Wonderful Year*, "Oh what an Earth-quake is the alteration of a State,"[11] and although he is speaking of the death of Queen Elizabeth I, early moderns understood exactly the wide-ranging, ecological change brought to full systems when wonders emerged and actants responded to the sudden, unpredictable, individually unwilled radical alteration in which all things change both internally and externally in relationship to one another.

There is a fearful wonder to such change wrought in the trembling toward rupture, and that trembling—as if something was trapped within and struggles to come to the surface—speaks to what Bennett calls "thing-power." Things offer up their own Titanic, Promethean "recalcitrance" (3) in challenge to human efforts to name and control them. The Titanic recalcitrance is, she says quoting Hent de Vries, "that which tends to loosen its ties to existing contexts" (qtd. in Bennett, 3), and, more importantly, in her own words, it is an event that is a human recognition that we should "name" as "the moment of independence (from subjectivity) possessed by things." We know these Titanically recalcitrant things, because under our newly attuned perception, they can show themselves more as they are, as "active, earthy, not-quite-human capaciousness (vibrant matter)" (3). This understanding of the material world also describes early modern experience of the *meteora* and specifically of the earthquake just at the moment it occurs and before one is compelled to interpret it by way of a sanctioned interpretive register. The earthquake can remind one, inconveniently, of the Titans that rumble beneath our individual, national, and religious foundations.

What is also at stake here, emerging as a through line in this volume but warranting book-length treatment unto itself, is the concept of spontaneous generation. Such generation is at once inexplicable, unwilled, and undeniable as an early modern sublunary experience. Toads, maggots, and other creatures were are among the products of this process that were commonly understood enough to gain regular reference. Perhaps currently the most notable literary representation of this spontaneous generation is Shakespeare's. When referring to the potency of the Nile and indirectly to Cleopatra, who is a force with whom the whole western world must grapple, Mark Antony and Lepidus agree regarding the remarkable fertility of Egypt:

Lepidus	You've strange serpents there?
Mark Antony	Ay, Lepidus.
Lepidus	Your serpent of Egypt is bred, now, of your mud by the operation of your sun; so is your crocodile.
Mark Antony	They are so.[12]

Here is generation brought about by the sun and mud, rather than seed and womb. It is truth: "They are so," but it is a truth that the *meteora* as a set runs counter to the laws of nature. This generation suggests a power beyond patriarchal reproduction and civilization. It calls to mind polytheistic conceptions of a generative earth who, without seed but with her own intent and passion, bears children. Spontaneous generation is also, we know now, a matter of scale and time. It is a phenomenon that ceases to exist once we introduce the microscope or time-lapse photography, for example. Without this technology or knowledge, however, one placed all phenomena philosophically into one of several longstanding paradigms. The full description and examples of these paradigms follow in chapter 1; here it is enough to say that all three methods for accounting for creation nevertheless found spontaneous generation problematic, because the only governing principle for these phenomena is that they appear with cause or meaning. They are accidents, monsters, and they seem then by their existence to figure in the early modern mind as a kind of wormhole or as dark matter, the positing of which itself admits incomplete knowledge related to the universe. More troubling is that spontaneous generation was more common in the witnessing of it than we might think, and it was more a matter of debate at the time, although not openly in scientific circles until the later seventeenth century when, as Lucinda Cole explains, notions of biogenesis and the reproduction from eggs were admitted as viable explanations.[13] Before then, biologically, worms and serpents, anger and erotic feeling appeared together as troublingly spontaneous in generation—along with all *meteora*, from hail to rainbow.

Earthquakes in particular exemplify the inexplicable, unwilled, forceful, and deeply troubling nature of spontaneously generated things. More profoundly still, the experience of the earthquake—its unique impact one thing to the next, its attendant effects, and all human and non-human responses to it—is just such a spontaneous generation traced out from its source with complexity and as an assemblage. Many of the components of the earthquake itself (—its location and magnitude, for example) like many of the components of the earthquake experience, will in a future occurence be similar, but never either identical, predictably emergent, interactive, or resolved. In a pre-scientific era, the earthquake and its wake had the power to disturb institutions and nations resembling recently in the U.S. the disruptions caused more recently by Hurricanes Katrina, Harvey, Irma, and Maria—the effects of which for many are ongoing. The early modern weather is not, however, our "weather"; nor is it the "weather" of our grandparents. As we take up the study of early modern meteorology, there is and will be room for surprise, for losing our footing, and for delight.

*

As the chapters that follow attest, early modern meteorology was an enticing and troubling subject, and its thoroughgoing relationship with physiology is evident in the frequency and complexity with which early modern writers across disciplines represent it. In chapter 1, "A Tale Put in for Pleasure," I provide description of the conceptual and experiential underpinnings of early modern meteorology. This includes examination of the terminology and of the theories early moderns employed as they felt, knew, and shared information regarding the physical realities of their universe. As much as they shared basic terminology, the theoretical explanatory systems at their disposal—what I will be calling the polytheistic, the materialist, and the Christian—varied greatly, allowing early moderns to chart a terrain for the representation of meteorological phenomena that is far more vast than that which has been available to us since that time. Chapter 2, "The Sneezing of the Earth," examines the reaction to the 1580 Dover Straits quake in popular, inexpensive printed texts. The over two dozen works entered in the Stationers' Register in the weeks and months following that earthquake include among them, in interesting order of entry and publication, an exhortatory religious pamphlet by soldier and writer Thomas Churchyard; one aiming to instruct readers in natural philosophy, written by physician, translator, and astrologer Thomas Twyne; a schedule of common prayer and homily issued by Queen Elizabeth I, her Privy Council, and the Church of England; and a translation of a Viennese treatise on the age of earth, with earthquakes as a symptom of earth's death by English Chronicles editor Abraham Fleming. Examination of the final and most intentionally outrageous of printed responses to the 1580 earthquake closes the chapter with the memory of a sneezing earth and an invitation for imaginative, playful responses to the quake. As a set, these pamphlets contain within them a significant portion of possible approaches to the textual representation of radical change, from Christian and materialist to polytheistic, from the predominantly imaginative, meant to elicit pleasure, to the expressly hortatory, meant to bring people to renewed faith through fear. Together, they more generally speak to the pervasive influence of meteorophysiological thinking and the dangers associated with spontaneous generation—even when that spontaneously generated thing is a sneeze or a laugh.

In chapter 3, "Much Enmoved, but Steadfast Still Persevered," the turn is decidedly literary in focus on the spontaneously generative and expansive meteorophysiological forces that fuel some of the most memorable monsters and heroes of Spenser's *Faerie Queene*. The first mention of Spenser's epic occurs in the letters of Harvey and Spenser treated in chapter 2. Ten years later, Spenser depicts in fantastic forms the gravest of meteorological challenges, when the elements of air and earth combine and make Titanic earth-shakers. These threats are meteorological, physiological, passionately charged, and apocalyptic. The only way to withstand them is to counter them not only spiritually with grace but meteorophysiologically with the staying power of resilience—a lesson taught to Redcrosse by Arthur. To conclude a study of Spenser's meteorology at this point, however, would be to deny examination of his most complex appropriation of such an event of trembling and of its answer in resilience. Spenser gives us Britomart, his knight of Chastity, whom he casts as mother earth erupting in a manner that seems initially spontaneous. Her virginal body begins to generate horrifying quakes that portend only rupture unless she or her nurse can determine their cause. With the help of Merlin, who teaches her of her destiny, Britomart learns by the end of book 3 to manage disruption by redirecting the forces within her, appropriating her own meteorophysiology to secure with steadfast perseverance the destiny she has glimpsed. Britomart emerges, ironically, as the most spontaneously eruptive body of the epic which is at once its stillest center. She is also in many ways the still center of this book project and anticipates the treatment of emotions, memory, and power in chapters 4 and 5.

In chapter 4, "Like an Overcharged Gun," the act of cursing in Shakespeare's first tetralogy becomes a political weapon against tyranny, but it is also a symptom of a dangerously ill sublunary body, able to extend its internal eruption beyond itself to other bodies within its compass. The fire that fueled Britomart's erotic physiological suffering and, in more directed form, her martial quest to find her beloved and by their union secure the future of Britain, is here used to fuel revenge and the War of the Roses. In the same decade in which Spenser published *The Faerie Queene*, Shakespeare created and staged *1–3 Henry VI* and *Richard III* (1591–92), offering Queen Margaret as the one so skilled in the act of cursing that it appears she can by way of memory and Galenic regimental techniques voluntarily bring about what would otherwise be the spontaneous generation of a curse. In the queen's highly effective efforts to exacerbate and then direct the forces animating radical material change,

Shakespeare represents the risk to humans who attempt to harness the sublunary, meteorophysiological forces to which they are also subject. Not everyone is a master of resilience, like Arthur and Britomart able to succeed in the endeavor.

In chapter 5, "These Signs Have Marked Me Extraordinary," I treat Shakespeare's representation of the experience of a literal earthquake in Shakespeare's plays *Romeo and Juliet* (1595); *1 Henry IV* (1596–97); and Shakespeare and Wilkins' *Pericles* (1607–08). In each play, a maternal figure—the nurse, the mother of Glendower, and Thaisa, respectively—experiences an earthquake as part of a rite of passage. Instructed by Louis Schwartz's illuminating *Milton and Maternal Mortality*, we might read the earthquake event that corresponds with birth as negative, as among the sufferings attendant upon these maternal bodies that God uses as a sign of Providential testing the better to result in heavenly reward.[14] This is not, however, what these plays reveal. Countering most treatments of earthquakes and birth in the period, and most treatments previously examined in this volume, the earthquake experiences in these plays are positive. They are associated with outright pleasures in the maternal body, and they stand as a poignant reminder that mutability also produces the spontaneous generation of joy.

Chapter 6, "These Earthquakes in Himself," brings the *Meteorology and Physiology* full circle with a return to a focus on prose and on the more personal nonfiction narrative. Anticipating his own death of an unknown feverous illness, John Donne imagines the microcosmic version of the sudden alteration that Thomas Dekker describes when Queen Elizabeth died: "Oh what an Earth-quake is the alteration of a State." In *Devotions Upon Emergent Occasions* (1624), Donne similarly cries out against the spontaneous changes he witnesses, but these are in his own body: "Is this the honour which man hath by being a *litle world*, That he hath these *earthquakes* in him selfe, sodaine shakings" (5). Donne's anxious self-scrutiny generates meteor-creating internal pressure that can unsettle the routine of the humorally constituted psychophysical body. Donne's internal quaking is also, like Dekker's "Earth-quake," a symptom of a very real concern that the former state will never be recovered. In Dekker's case, that former state is, already nostalgically, the golden age under Elizabeth I, mother of England. In Donne's case, the former state will one day, postmortem, be that of his body and soul, once knit together in his earthly life. The Christian promise of resurrection should result in a reconstitution of that union after death, at least for the faithful, but Donne is deeply troubled by "these *earthquakes* . . . sodaine shakings" that suggest a material rupture so radical that no fathomable recovery will be satisfying even if it is possible. Perhaps as he fears, death will show that we humans have only ever been *meteora*, spontaneously generated, *sans* cause, *sans* signification, soon to dissolve, in the words of Shakespeare's Jacques, "sans everything."[15]

This is the threat of the earthquake in this period. This is the threat of the trembling meteorophysiological body that is always also an assemblage: it can devolve—spontaneously without will or cause—into pieces impossible to recollect, into pieces that will become actants in some *other* assemblage. This the earthquake's appeal in textual representation, its appeal as a subject in inexpensive, popular print pamphlets; on London stages; in the epic; and in personal devotions alike. Representations of metamorphosis undeniably stir the imagination and incite new thinking about the composition of the world and the place of humans within it—new thinking that makes an impression on those who entertain it. Ovid knew this. Playing the Titan to Rome's Caesar, he rumbled, at least in literary form, undermining Rome's Virgilian historical narrative of itself as preeminent among civilizations in of world history. Harvey, Spenser, Shakespeare, and Donne admired and imitated this Ovidian rumbling, even when doing so risked disquieting aftershock. The chapters that follow highlight the wondrously complex effects of their admiration that play out in an imaginative expanse of textual representation that is no longer as large. By the latter half of the eighteenth century, with the rise of science that would see the formation of geology and eventually, in the nineteenth century, seismology as disciplines, these matters—earthquakes, Titans, mother earth, and the radical Ovidian transformation of bodies—had begun to diminish in emotional and imaginative capital, and then in memory. But prior to this, we encounter in England an impressive and intentionally delightful range of descriptive registers available to capture the experience of all meteorophysiological bodies trembling toward rupture and rapture, fearful and marvelous.

Notes

1. Shigehisa Kuriyama, *The Expressiveness of the Body and the Divergence of Greek and Chinese Medicine* (New York: Zone Books, 1999), 162.
2. Katharine A. Craik and Tanya Pollard, *Shakespearean Sensations: Experiencing Literature in Early Modern England* (Cambridge: Cambridge University Press, 2013), 7–8; See also, following their note, Gail Kern Paster, *Humoring the Body: Emotions and the Shakespearean Stage* (Chicago and London: University of Chicago Press, 2004), 6; Gail Kern Paster, Katherine Rowe, and Mary Floyd-Wilson, *Reading the Early Modern Passions: Essays in the Cultural History of Emotion* (Philadelphia: University of Pennsylvania Press, 2004), 16.
3. Paster, *Humoring the Body*; Michael C. Schoenfeldt, *Bodies and Selves in Early Modern England: Physiology and Inwardness in Spenser, Shakespeare, Herbert, and Milton*, Cambridge Studies in Renaissance Literature and Culture 34 (Cambridge, UK: Cambridge University Press, 1999).
4. Please see essential work in this area of human embeddedness in the aptly titled *The Indistinct Human in Renaissance Literature*, Early Modern Cultural Studies 1500–1700, eds. Jean E. Feerick and Vin Nardizzi (New York: Palgrave Macmillan, 2002).

5. Christopher Marlowe, "*Tamburlaine the Great: Part One*," in Joseph S. Cunningham and Eithne Henson, editors, *Tamburlaine the Great* (Manchester: Manchester University Press, 1998), 1.2.47–51.
6. William Shakespeare, "*The Tempest*," in Virginia M. Vaughan and Alden T. Vaughan, editors, *The Arden Shakespeare*, Third Series (London: Arden Shakespeare, 1999; London: Bloomsbury Publishing, 2011), 1.2.401.
7. Shakespeare, "*King Henry V*," in T.W. Craik, editor, *The Arden Shakespeare*, The Third Series (New York and London: Routledge, 1991), 3.1.6–17. All citations for this play are from this edition and in this volume will be cited in the text by act, scene, and line.
8. Paster, *Humoring the Body*; Schoenfeldt, *Bodies and Selves in Early Modern England*. On new materialist readings of the gendered body, interesting extensions of the work formerly mentioned, see Diana Coole, "The Inertia of Matter and the Generativity of Flesh," in S. Frost, editor, *New Materialisms: Ontology, Agency, and Politics* (Durham, NC: Duke University Press, 2010), 92–115.
9. Gilles Deleuze and Félix Guattari, *A Thousand Plateaus: Capitalism and Schizophrenia* (Minneapolis: University of Minnesota Press, 1987), 8, 88; Jane Bennett, *Vibrant Matter: A Political Ecology of Things* (Durham: Duke University Press, 2010), viii, xvii. See also Manual DeLanda for help, as from Bennett, in the manageable use of Deleuze and Guattari's concept: Manuel DeLanda, *A New Philosophy of Society: Assemblage Theory and Social Complexity* (London: Continuum, 2006); and Manuel DeLanda, *Assemblage Theory* (Edinburgh: Edinburgh University Press, 2016).
10. OED s.v. mine, noun, 3.a. "Formerly: a subterranean passage dug under an enemy position, esp. the wall of a besieged fortress, in order to gain entrance or to bring about its collapse. In later use: such a passage in which an explosive is placed. Also [occas.]: the explosive charge so placed. Now *rare* [*hist.* in later use]."
11. *The wonderfull yeare. 1603 Wherein is shewed the picture of London, lying sicke of the plague. At the ende of all (like a mery epilogue to a dull play) certaine tales are cut out in sundry fashions, of purpose to shorten the lives of long winters nights, that lye watching in the darke for us*. London: Printed by Thomas Creede, and are to be solde in Saint Donstones Church-yarde in Fleet-streete [by N. Ling, J. Smethwick, and J. Browne] (1603), sig. B2v.
12. William Shakespeare, "*Antony and Cleopatra*," in John Wilders, editor, *The Arden Shakespeare*, Third Series (1995: London: Arden Shakespeare, 2006), 2.7.24–27.
13. Lucinda Cole, *Imperfect Creatures: Vermin, Literature, and the Sciences of Life, 1600–1740* (Ann Arbor: University of Michigan Press, 2016), 13, 28–29, 33. See Wolfram Schmidgen, *Exquisite Mixture: The Virtues of Impurity in Early Modern England* (Philadelphia: University of Pennsylvania Press, 2013), 39, 51, 75.
14. Louis Schwartz, *Milton and Maternal Mortality* (Cambridge: Cambridge University Press, 2009).
15. Shakespeare, "*As You Like It*," in Richard Proudfoot, Ann Thompson, and David Scott Kastan, editors, *The Arden Shakespeare Complete Works* (London: Arden Shakespeare, 2002), 2.7.167.

1 A Tale Put in for Pleasure

Meteorology and Physiology in Early Modern Culture: Earthquakes, Human Identity, and Textual Representation treats the perils and pleasures of the written expression of the earthquake experience and its physiological equivalents. The philosophical underpinnings of these provocative textual representations of meteorophysiological wonder are as complex and varied as the representations themselves, and they are also quite removed from our current habits of thought about these subjects, informed as we are by centuries of the scientific method and belief in a heliocentric rather than geocentric universe. Given this fact and, on the other hand, the widespread, enduring, and popular early modern interest in meteorology, it is worth naming explicitly some of the reasons that current literary scholars and historians have only recently turned their attention to the subject that seems so in need of and able to reward attention.[1] Katherine Cox accounts for the relative lack of scholarship on early modern meteorology as she treats "the air" as a phenomenon of the weather in Milton's *Paradise Lost*: "Scholarship has underestimated the pertinence of meteorology to the epic possibly because modern notions of the weather exclude phenomena that in Milton's era were deemed meteorological . . . such as mist, dew, and rain . . . comets and shooting stars,"[2] as well as earthquakes, tsunamis, and volcanos. My own sense of the situation is similar: we have "underestimated" the importance of meteorology in the premodern period, and this may be due to our thinking that "the weather" is a mundane subject (as indeed it so sounds if one thinks generally of a paper on the weather in *Hamlet*, for example); it may also be due to own relatively limited experiences with and knowledge of the weather and of meteorology, which is the result of our dependence on television news and internet satellite information rather than on experiential data we might examine in order to determine how the weather might impact us any given day. It is also the case that early modern writing about meteorology is not easily identified or valued, and it is for these reasons together that in this chapter I provide something of a primer on the essential meteorological theory and terminology widely and regularly applied by early moderns across classes, disciplines, and media.

*

Early moderns inherited the basic terminology for and understanding of meteorology that Aristotle codified in his physical works. The title for his *Meteorology* derives from the ancient Greek word *μετέωρα*, meaning, according to the Oxford English Dictionary, things "lofty" or "raised."[3] We might take this to mean things raised above the earth or emitted from it, as in a breath, making it all the more difficult for us to understand now how earthquakes, rivers, gems, and metals were also considered among *meteora*. They were all of them produced by processes involved in a kind of universal respiration and digestion: the constant exchanges among the four elements as well as their conversions into each other. Meteorology therefore subsumed fields we now call geology and seismology as well as some things we consider the subjects of astronomy. This was also a field in which subjects were inaccessible and changeable, frustrating efforts at scientific quantification of the sort one might apply in other disciplines, such as those supplying subjects more accessible to the senses or easier to quantify.

By the time of Queen Elizabeth I, several cosmological paradigms appear at once in the service of accounting for these elusive phenomena. It is as if it required a number of fully developed paradigms, each stretched to their imaginative limits, to offer the range of possible interpretations and other responses that the variety and uncertain nature of meteorological phenomena demanded. Dominant among these paradigms are the three I take as focus in this project, because they appear in nearly every treatment of the physical world available in the period. I call these paradigms the "polytheistic," first codified by Hesiod and which featured those Titans; the "materialist," by Aristotle; and the "Christian," of the Bible.[4] All three paradigms were employed simultaneously, but the "materialist" was on swift ascent, generally accepted as a valuable method for describing natural events, as indicated by the terminology for these events being taken from Aristotle, including the very specific concept of "exhalations," which is his creation. Christians aimed, however, to assert atop the materialist paradigm a Christian creator God who put the Aristotelian world of nature in place and who, they sought to assert, did so to purpose bending the rules of nature as he pleased. Every meteorophysiological representation in the period showcases beliefs and language derived from these two paradigms. Always also present in meteorophysiological representation, though in less thoroughgoing form, are features of the polytheistic paradigm which posits the random passion of a sentient, generative mother earth as fuel for all creation. Earth in this paradigm is the mover of all sublunary things, and, to underscore, Earth's here is a motion born of passion that is largely life-producing and therefore positive.

A supplement more often than a lead explanatory framework in the period with respect to what might be considered truth, the polytheistic model nevertheless was with the materialist model dominant in imaginative representations of meteorophysiological wonder. These two models together and alone gave writers wide range for the construction of populated worlds—from historical recreations of the War of the Roses in Shakespeare's plays to the land of Faerie in Spenser's epic. It is striking, in fact, the degree to which literature of the time leaves Christian meteorophysiology in the background to highlight theories from the other two cosmological paradigms. A polytheistic interpretation of an earthquake as mother earth's Titan children rumbling within her is more interesting than the materialist version of wind trapped in earth, shaking earth as it forces its way out, and both of these explanations, which work so well together to expand description of a quake, are more interesting than the biblical account of quakes as God's enactment of judgment on sinners. The latter explanation was taken as an overriding given with respect to "truth," but received little airtime, except, as we will see, when an event such as the 1580 earthquake demanded its reinforcement. Even then, in Christian pamphlets on monstrous births and earthquakes, and in the 1580 quake responses, writers struggled to silence the power of the polytheistic and materialist views of creation.

*

Because we encounter aspects of all three paradigms regularly in nearly all texts from the period, the balance of this chapter treats the essential features and major explanatory strengths and weaknesses of the three primary interpretive paradigms for meteorology that were available to early moderns. The first record we have in the western tradition for the interpretation of change in the weather posits a polytheistic cosmology with a generative mother earth as its primary creator. Her *meteora* are her progeny. In the earliest codified account of this paradigm, the *Theogony* by Hesiod (c. 700 B.C.), Earth is a passionate and stable creator:

> Verily at the first Chaos came to be, but next wide-bosomed Earth, the ever-sure foundation of all the deathless ones who hold the peaks of snowy Olympus, and dim Tartarus in the depth of the wide-pathed Earth, and Eros (Love), fairest among the deathless gods, who unnerves the limbs and overcomes the mind and wise counsels of all gods and all men within them. From Chaos came forth Erebus and black Night; but of Night were born Aether and Day, whom she conceived and bare from union in love with Erebus. And Earth first bare starry Heaven, equal to herself, to cover her on every side, and to be an ever-sure abiding-place for the blessed gods. And she brought forth long Hills, graceful haunts of the goddess-Nymphs who dwell

> amongst the glens of the hills. She bare also the fruitless deep with his raging swell, Pontus, without sweet union of love. But afterwards she lay with Heaven and bare deep-swirling Oceanus, Coeus and Crius and Hyperion and Iapetus, Theia and Rhea, Themis and Mnemosyne and gold-crowned Phoebe and lovely Tethys.[5]

Not only is Hesiod's earth generative, "she" is the "ever-sure foundation" of all things, even of "all the deathless ones," the gods themselves. And among her children—meteoric issues from her body—are the Titans and the Olympian gods, including Prometheus and Zeus, the son of Cronos and thus her grandchild. Hesiod's earth is also literally mother of all *meteora*, including, by premodern categorization, the ocean and rivers. Each of these children bear names and autonomous emotions, wills, and generative powers. The gods named last in the quoted passage are the first Titans, from whom Prometheus and then humans descended. In this codified polytheistic conception of the beginning of the universe, mother earth is the origin of all sublunary things human and non-human. Her reason for generation is passion—nothing more than this unstated but evident imperative.

For centuries, this view of creation, of an *anima mundi* as an often autonomous source of generation in the universe, dominated. In Rome, Ovid offered Hesiod's polytheistic creation account infused with Lucretian atomism in an effort to counter a monolithic version of history as told by Virgil in his *Aeneid*.[6] Ovid's tales of random change that was the product of equally random creative passions of the gods challenged Virgil's tale of Rome's national destiny; Ovid would rumble as a Titan against Virgil's tidy story and, through the concept of metamorphosis, serving metonymically for all such sub-narrative rumblings in texts and lives, nations and philosophies. Ovid's *Metamorphoses* supplied an enticing view of bodies and of history that quickly expanded the representational registers of those who encountered the stories immediately and for centuries thereafter.[7] As we know, for example, Ovid's work influenced that of artists across nations and media, from Botticelli and Titian to Petrarch and Chaucer. In Todd Borlik's words, specifically with the polytheistic earth in mind, "[t]he concept of the *anima mundi* . . . came down to the Renaissance with a sterling pedigree, championed by some of the greatest cultural figures of antiquity, such as Plato, Pliny, Virgil, Cicero, Hermes Trismegistus, and [ ] Ovid."[8] In early modern England more specifically, Ovid spoke persuasively to writers who sought power over the imaginations of their readers. The polytheistic cosmological paradigm and its unusual, disruptive accounts of creation, change, and destruction supplied that need. Most of the writers from the period that we celebrate today held Ovid in high regard for this reason, finding his great work a source for inspiration, expanding their imaginative terrain by allowing for a kind of narrative plasticity. Spenser, Shakespeare, and Donne are among those writers and thinkers, all of them benefitting from their early grammar school exposure to the anything-but-tidy *Metamorphoses*. As Lynn Enterline explains, "the institutional and discursive practices of England's rhetorical culture—specifically, the Latin curriculum of the humanist grammar school—[ ] gave Ovid's texts such a crucial place in Elizabethan poetry and drama."[9]

*

It was Arthur Golding's 1567 translation in particular that extended further Ovid's cultural reach. It was, as Oakley-Brown reminds us, "one of the most influential translations" ever for Ovid's *Metamorphoses*. Golding's translation was also, she points out, an intentional effort to "domesticate[ ]" Ovid for a Protestant audiences, and it is worth turning here from treatment of the polytheistic paradigm to the most interesting of early modern efforts simultaneously to share and to mute it (12). A pronounced case of Golding's domestication effort is his preface to his translation of Ovid, which comes in the form of a verse epistle dedication to Robert Dudley, Earl of Leicester. In this epistle, Golding includes among other things a biblical interpretation of some of the decidedly non-Christian content to follow in his Ovid. He is obviously troubled by Ovid's creation of humans in book 1: "*natus homo est, sive huc divino semine fecit/ille opifex rerum, mundi melioris origo,/sive recens tellus seductaque nuper ab alto/aethere cognati retinebat semina caeli./quam satus Iapeto, mixtam pluvialibus undis,/finxit in effigiem moderantum cuncta deorum.*" By way of Frank J. Miller's Loeb Classical Library translation, this might read:

> Then man was born: whether the god who made all else, designing a more perfect world, made man of his own divine substance, or whether the new earth, but lately drawn away from heavenly ether, retained still some elements of its kindred sky—that earth which the son of Iapetus mixed with fresh, running water, and moulded into the form of the all-controlling gods.[10]

Golding renders these lines in two ways, in his translation proper, directly below, and in his epistle, which will follow thereafter. In the translation,

> Then eyther he that made the worlde, and things in order set,
> Of heavenly séede engendred Man: or else the earth as yet
> Yong, lustie, fresh, and in hir floures, and parted from the skie,
> But late before, the séede thereof as yet held inwardlie.
> The which *Prometheus* tempring straight with water of the spring,
> Did make in likenesse to the Gods that governe everie thing.

In the similar but not identical epistle rendering of these lines, he offers

> Then eyther he that made the world and things in order set,
> Of heavenly seede engendred man: or else the earth as yet
> But late before, the seedes thereof as yit hild inwardly,
> The which Prometheus tempring streyght with water of the spring,
> Did make in likenesse to the Goddes that governe every thing.[11]

The subtle changes, between Ovid and Golding and between Golding's two versions in the same text are worth attention for many reasons, but most immediately I focus on what Golding does with Ovid's earth. Golding's augmentation of Ovid's "recens tellus" or "fresh/new earth" (translation mine), making earth "as yet/Yong, lustie, fresh, and in hir floures" is titillating and diminishes her power at the same time. It turns her into a girl, ready to be taken—a change that elides any individual creative power Ovid attributes to his "recens tellus" and nearly power that an explicitly Hesiodic Gaia held.

More interesting are these lines below, following on the heels of the last line cited previously from Golding's preface:

> What other thing meenes Ovid heere by terme of heavenly seede,
> Than mans immortall sowle, which is divine, and commes in deede
> From heaven, and was inspyrde by God, as Moyses sheweth playne?
> And whereas of Prometheus he seemes too adde a vayne
> Devyce, as though he ment that he had formed man of clay,
> Although it bee a tale put in for pleasure by the way:
> Yit by thinterpretation of the name we well may gather,
> He did include a misterie and secret meening rather.

Golding comes in immediately over Ovid's depiction of creation possibly at the hand of Prometheus, seeming to correct an implication Ovid really does not make, that the Titan is "a vayne/Devyce," and that "it bee" only "a tale put in for pleasure, by the way." He does this in order to offer a more credible account for Ovid's use of Prometheus, but in the process, he implies nevertheless than Ovid is just telling tall tales, nothing to be taken seriously. In the account we are to take to heart, he hints at the "misterie and secret meening" he has gleaned from Ovid. It is—mystery and meaning, he reveals next, that suits a Christian interpretation for creation:

> This woord Prometheus signifies a person sage and wyse,
> Of great foresyght, who headily will nothing enterpryse.
> It was the name of one that first did images invent:

Of whom the Poets doo report that hee too heaven up went,
And there stole fyre, through which he made his images alyve:
And therfore that he formed men the Paynims did contryve.
Now when the Poet red perchaunce that God almyghty by
His providence and by his woord (which everlastingly
Is ay his wisdome) made the world, and also man to beare
His image, and too bee the lord of all the things that were
Erst made, and that he shaped him of earth or slymy clay:
Hee tooke occasion in the way of fabling for too say
That wyse Prometheus tempring earth with water of the spring,
Did forme it lyke the Gods above that governe every thing.
Thus may Prometheus seeme too bee theternall woord of God,
His wisdom, and his providence which formed man of clod.
(sig. B2v–r)

Goldings spends many words to account for Ovid's suggestion, following Hesiod, that Prometheus had a hand in creating humans. Prometheus could not, for Golding, be such a creator, but he can stand as a sign for creation as a process, as "a person sage and wyse" who "first did images invent" like a poet and who, then, is not a God unto himself but is a sign for "the eternall word of God"—all "in the way of fabling." Any power Prometheus has, Golding is at pains to show, is fictional and figurative. Moreover, "earth" here appears only to be one of the four elements, inert, a substance mixed rather than the sentient being suggested by Ovid's words.

Golding's translation and his prefatory explanation appear in this context together as efforts both to highlight his own authority and skill, meriting patronage, and to exercise productively the general early modern anxiety felt in response to all non-Christian accounts of creation, change, and destruction. In this case "non-Christian," it should be noted, does not include something like Renaissance Neoplatonism, which like Aquinas' mixing of Aristotelian materialism and Christian theology, was a blend acceptable to most Christians as a form of Christian philosophy. We know from the number of early moderns who turned to Golding's Ovid to fuel their tales of metamorphic alteration, however, that Golding's translation helped turn the volume up on rather than diminish the polytheistic notes strongly sounded by his source. Ovid's tales are irrational, passionate, and not at all teleological from a Christian perspective, each one a celebration of change that is always playfully both painful and pleasurable and is never certain in its origin, form, or ending, sans prescription. Just so, the tales are also each a Titan, an earthquake, like early modern *meteora* generally in their disruptive, memorable, and, in Bennett's words, "recalcitrant[ly]" untamable "thing-power" (3). This is the power of stories of the "wide-bosomed Earth" and of Titans like

Prometheus who erupted irrepressibly in imaginations and in memories, taking many forms. "Put in for pleasure," certainly, such stories are nevertheless never merely "vanye/Devyce"s.

*

The power of the polytheistic conception of the universe was enormous, dominant in the west well before its codification by Hesiod in the eighth century B.C. Alternative theories had been mounted by the pre-Socratics and Plato, but by the time Aristotle was writing in the fourth century B.C., centuries after Hesiod, the polytheistic system had gone without fully successful challenge. Aristotle succeeded in his attempt to take the passion out of the system and reduce its mystery. In his physical lectures, delivered c. 350 B.C, Aristotle offers a rational, non-agential account of all physical phenomena in the universe to replace the polytheistic account with its mother earth as passionate, sentient prime mover. He codified what I am calling the "materialist" cosmological paradigm—a paradigm that endured well beyond its origins, just as would the polytheistic paradigm it sought to replace. Influential natural philosophers including Seneca and Pliny the Elder carried forward Aristotle's theory, and Arab translators in the Middle Ages saw that it would survive church efforts to silence it, and it is this theory that would eventually remain over the others as most viable.

Part of the viability of Aristotle's theory of universal matter and motion is its focus on the identification of the distinctive qualities of the parts of the universe at least as much as on contemplation of unifying patterns among them. His contribution to natural philosophy is to advance convincingly the proposition that all change is naturally, unintentionally occurring because it is generated by the properties of its own matter. It is not, as in polytheistic (and later Christian) teaching, the creation of a sentient and/or passionate individual. For Aristotle, the matter of the universe coheres into things, persists, and resolves back into the four elements—all in adherence to observable physical laws that humans can identify and by which often they can predict outcomes for things (the egg precedes the chicken; the accumulation of clouds precedes a thunderbolt; the lack of regular breathing precedes an angry outburst; it is not up to gods or goddesses, because it is a natural process). With this reasoning, superstitious rituals performed to appease the gods out of fear or habit are unnecessary.

In his physical lectures, Aristotle more specifically sought to distinguish a range of sublunary phenomena from those in other sublunary regions of the universe—astronomical from meteorological from biological. Aristotle begins by reviewing the largest category of matter and motion in his lectures called the *Physics*. This is the first of his five-part series of lectures in which the *Meteorologica* figures third, in the very middle as a lynchpin. It is in this middle book that Aristotle summarizes the general sublunary motions that govern the many changes occurring in the region below the moon:

> We have already laid down that there is one principle which makes up the nature of the bodies that move in a circle, and besides this four bodies owing their existence to the four principles [hot, cold, moist, dry], the motion of these latter bodies being of two kinds: either from the centre or to the centre. These four bodies are fire, air, water, earth. Fire occupies the highest place among them all, earth the lowest, and two elements correspond to these in their relation to one another, air being nearest to fire, water to earth. The whole world surrounding the earth, then, the affections of which are our subject, is made up of these bodies.[12]

All things in this region—the comets, earthquakes, volcanoes, tidal waves, winds, fires, rocks, and springs, like humans, animals, plants, trees, and metals—are made of and animated by these elements and their motions. As we currently might say that humans and all things are constituted materially of stardust or, more precisely of 12 of the elements of the elemental table, they could make a more substantial claim: all things in the sublunary region—each and every thing—was materially composed of four elements—what we now call "the four the elements" of earth, water, air, and fire.[13] In this lecture, Aristotle shared in detail as well the dominant qualities, motions, and locations of each element: hot, cold, moist, and dry; up, down, circular; earth center to crust, water above earth, air above water, fire above air extending to the sphere of the moon. These qualities and motions he saw as constitutive of sublunary matter such that the dominant element composing any thing would help determine that thing's trajectory and habitation.

The opening lines of his *Meteorology* underscore the aims for his physical lectures as a whole, including the interest in distinction as much unification among the things of the universe:

> We have already discussed the first causes of nature, and all natural motion, also the stars ordered in the motion of the heavens, and the corporeal elements—enumerating and specifying them and showing how they change into one another—and becoming and perishing in general. There remains for consideration a part of this inquiry which all our predecessors called meteorology. It is concerned with events that are natural, though their order is less perfect than that of the first of the elements of bodies. They take place in the region nearest to the motion of the stars. Such are the milky way, and comets, and the movements of meteors. It studies also all the affections we may call common to air and water, and the kinds and parts of the earth and

> the affections of its parts. These throw light on the causes of winds and earthquakes and all the consequences of their motions.
>
> (554; 1.1.338a20–339a9)

In addition to being a useful outline of his *Meteorology*, this statement is also the single most important overview that Aristotle provides for his entire examination of the universe, and it is worth note that such an overview appears here rather than in the opening for the *Physics* or *On the Heavens*. In this passage, Aristotle gives account of his books on the physical sciences and places his *Meteorology* central among them, as a key to the other studies. As noted previously, he sees the *meteora* as the linchpin-like set of phenomena between the materials and motions of the heavenly sphere and those of sublunary biotic creatures.[14]

This introduction also hints at the trouble with the subjects of study—with the difficulty in observing them to begin with let alone with placing them within a unifying structure allowing for their contemplation as a comfortable component of a harmonious universal system. In his words:

> Of these things some puzzle us, while others admit of explanation in some degree. Further, the inquiry is concerned with the falling of thunderbolts and with whirlwinds and fire-winds, and further, the recurrent affections produced in these same bodies by concretion. When the inquiry into these matters is concluded, let us consider what account we can give, in accordance with the method we have followed, of animals and plants, both generally and in detail. When that has been done we may say that the whole of our original undertaking will have been carried out.
>
> (554; 1.1.338a20–339a9)

Some of them "puzzle," and they do so with no resolution. For the "others" that "admit of explanation," still the best one can do is settle for an answer "in some degree." The subjects of study in this region are so changeable, it is difficult to observe them in the first place and nearly impossible to place them in a reasonable continuum with the other things of the universe, superlunary or sublunary.

Aristotle did what he could generally then to group these things into categories by their dominant elements (fiery, airy, watery, and earthy), but this too caused a problem. As shifting, in-between things, *meteora* could not cohere at any stage enough to allow humans to identify confidently their forms or sometimes even dominant elements. According to Malcolm Wilson, this is part of the reason Aristotle settled on an unusual fix by creating "exhalations"—one dry and inhabiting upper regions, the other vaporous and closer to earth, to help close the gaps between theory and observation. As Aristotle explains of these exhalations:

> When the sun warms the earth the exhalation which takes place is necessarily of two kinds, not of one only as some thing. One kind is rather of the nature of vapor, the other of the nature of a windy exhalation. That which rise from the moisture contained in earth and on its surface is vapor, while that rising from the earth itself, which is dry, is like smoke. Of these the windy exhalation, being warm, rises above the moister vapour, which is heavy and sinks below the other. (559; 1.341b6–12)

These exhalations were separate from the elements but interacted with them, the better to account for the distinction Aristotle observed between the *meteora* and the purer and more stable heavenly entities and between the *meteora* and the more complex, less stable but locally specific and entirely individuated biotic creatures populating earth.[15] Yet the *meteora* were still always ultimately unpredictable in their local manifestations, in spite of all of Aristotle's efforts to identify their materials and motions. Stated in positive terms each manifestation of meteor was entirely unique in every portion of its existence.[16]

This lack of predictability and certain distinctiveness to an individual meant that any effort to associate with them all four types of Aristotelian causality was impossible. Aristotle had adhered in most cases to the notion that the things of the universe have four types of cause: material, efficient, formal, and final. Craig Martin explains in *Renaissance Meteorology* that the need to associate all things with the four "causes was central to natural philosophy and a legacy of the influence of Aristotle's *Libra naturales* from the Middle Ages to the middle of the seventeenth century. In Aristotelian natural philosophy, causation is identical to explanation."[17] Etienne Gilson explains in greater detail the importance of causation. Although his own argument leads to the conclusion that the natural world contains within it its own teleology (which then can be suited to Christian one), his explanation here helps us to account for the stakes involved if one is unable to claim for all things the common four-fold causation; for this reason, I quote it at some length:

> Different kinds of causes are at work in nature: the material, the formal, the efficient [*le moteur*], and the final. All whose structure is homogenous can be explained by the efficient cause, which Aristotle frequently calls "the point of origin of motion." Heterogeneous parts require addition, for their explication, another kind of cause, that which we call today "the final cause", and which Aristotle calls simply "the end" (*telos*), the "in view of which" (*to ou eneka*), the

> "why" (*dia ti*). . . . If there is in the real a principle of unity—substance, for example—it is necessary that the four kinds of causes be able to return, in one manner or other, to this principle; a cause of any kind whatsoever is such only through it. . . . Aristotle expressly contradicts the Platonist notion which makes of life a simple source of motion, as if one single and identical thing could be motive force and thing moved at the same time and in the same way. It suffices to see an animal move about to ascertain that the parts which move take their point of departure from the fixed and the immobile. All living operations, all the growth of plants or animals, involve and require the differentiation of certain parts capable of acting one on another. . . . The ability of a living being to move itself, even though it be only to assimilate and grow, involves therefore the organization of the heterogeneous parts of which it is composed. This is why one says of living bodies that they are organisms or that living matter is organic [*organsisee*]. The finalism of Aristotle is an attempt to give a reason for the very existence of this organization.[18]

Gilson roots his discussion in evidence from Aristotle's *On the Parts of Animals* and his own observations of animals and readings in animal biology. As a practicing Thomist himself, and like most Christians, he wants to assign final cause to all things of nature thus able to posit God as their origin and to point to the larger, Christian meaning of every natural thing. His focus on biological creatures allows him to do so. What he neglects, conveniently or intentionally, is the category for which Aristotle could assign no formal or final cause.

The one set of phenomena that defies such categorization for Aristotle, and would for Gilson and others had they faced it squarely, perhaps even by way of Thomist synthesis, is the largest subset of "accident," the *meteora*. As shifting, in-between things that could not cohere, that shows no organization, it was impossible to identify all four of the causes for *meteora*. One could assign them material cause, because they were constituted materially of the elements and exhalations, and the Aristotelian efficient cause was the rotation of the spheres that churned these elements and exhalations. The matter and motion was in common among *meteora* and all sublunary things, but the individual manifestation of each could be strikingly unique—the triple rainbow, the sunshower the 100-year flood, the hail as large as apples. When is an earthquake, thunderbolt, or rainbow at the maturity of its form if it was always changing? What was the life arc of a snowflake or a thunderclap, with each so different and so quickly becoming some other form of meteor or entirely dissipating? The snowflake becomes a drop of water; the thunderclap ends in silence; moreover, the snowflake too be could as easily not cohere, not form, remain some partial water-ice thing not distinctly a snowflake.[19] Furthermore, how could one categorize an event like an earthquake that merged air and earth but that behaved in ways more like thunder than like other *meteora* closer to earth?[20] Aristotle chose not to assign final or formal causes to *meteora*. It was this observation and choice that would contribute to Christian discomfort with and in some cases rejection of his cosmological theory.

*

In England under Elizabeth I, there was ample and growing interest in the materialist position leading in some circles concern about this increase in popularity of materialist explanations for events. Religious writers in particular wrote of concern that people would start to follow materialist conclusions and find all wonders to be entirely natural in cause and content. Perhaps they would find human life itself to be only so much "sound and fury/Signifying nothing," all things only so many meteors.[21] All things might be, by this thinking, entirely the product of spontaneous generation. Thus did some fear a path to atheism set out within materialist interpretations of wondrous events. In yet other words, to explain the weightiness of the situation, the phenomenon of a meteor—tempest, rainbow, or comet alike—might be an "accident." As Michael Witmore reminds us, Aristotle refused to consider accidents as having the causality one could determine for other sublunary matter and motion.[22]Accidents, Aristotle found, simply lacked compositional coherence and duration; unable, in other words, to hold their forms, or even to hold still, long enough to allow for identification of their parts, beginning, middle, end, inner, outer, first, last. One could not, then, assign useful meaning to such events. Aristotle found most *meteora* fell into this category, and among accidents, *meteora* were the largest subset of entities in the sublunary sphere thus unable signify beyond themselves. In thus "signifying nothing," accidents, and their large subset of the *meteora* were for him not worth the trouble to try to turn them into stable signs. As discussed previously, this particular theory of Aristotelian meteorology proved most troubling for Christians encountering it, because it problematized biblical claims regarding the significance of a post-flood rainbow of promise, of Noah's flood, of the parting of the Red Sea, and of other wondrous events included there as signs of something greater than themselves.

Whatever problems it contained for Christians, however, Aristotle's materialist view of meteorological phenomena was a powerful one, with its own recalcitrant truth that spoke through all efforts to mute it. In England, during Queen Elizabeth I's reign, it gained another strong advocate for its worth. Ironically, perhaps, this strong voice for materialist meteorology came largely from theologian and fellow and later master of Pembroke College, Cambridge, William Fulke (1536/7–1589). Fulke emerged as the leading writer on the subject of materialist meteorological theory in part due to his larger anti-prognostication campaign.[23] This campaign was part of an increasing effort on the part of Protestants to separate themselves further from Catholics on issues they saw as most easily identified. For example, debating whether or not Christians should either fast as part of a program of Protestant penance or take the Eucharist as the real presence of Christ, was fraught with difficulty, due to centuries of widespread Catholic use, but, as Stuart Clark explains, some issues were easier to tackle, such as challenging the Catholic link between heretics and witches. As Mary Thomas Crane puts it succinctly, "Protestants went farther than any other European thinkers in expanding the number of phenomena considered to be purely natural and reducing the role of the supernatural."[24] Protestants nevertheless undertook careful examination of and debate over these phenomena, as exemplified by Fulke's work and that of the earthquake pamphlet writers of 1580. To suggest that the age of miracles was dead in some sort of monolithic fashion—as many literary scholars have done for the post-Reformation period and often citing Shakespeare's Canterbury in *Henry V* ("for miracles are ceased" [1.1.67])—is to deny the serious nature of what instead was intense but conflicting belief in the supernatural among Protestants. For Protestants, labeled heretics by Catholics, it was essential to distinguish natural events—with no specific message from God—from supernatural, considered direct communication from God. Moreover, centuries of Catholic interpretation of the supernatural had supported Catholic Church and not always decidedly biblical practice. The effort to assess and pronounce upon radical meteorophysiological events was thus intensified in this post- and counter-Reform era.

In England, William Fulke was among those leading this charge. Treating Fulke's efforts at length and reading them in light of related efforts to wrest authority over miracles and the supernatural generally from Catholics, Crane finds Fulke "perhaps more dedicated than anyone else in sixteenth century England to explaining all phenomena in terms of Aristotelian naturalism" (71), and this includes John Dee and Leonard Digges, who had taken up and to some extent popularized the subject of meteorology prior to Fulke.[25] By the 1570s, Fulke had found favor under Robert Dudley the Earl of Leicester, confidante to the queen and Privy Council member.[26] Fulke gained recognition as a staunch defender of the Church of England against Catholicism for the reasons suggested previously. Specifically, the Church of England was walking the line between Catholic and Calvinist readings of biblical prescription; to keep its practitioners along the full spectrum of confessions happy enough to avoid rebellion, the church relied upon men like Fulke to help explain that what Catholics or Puritans might want to show as miraculous signs of God's disapproval with England were instead signs of nature working quite naturally, thereby denying efforts to use those events for propagandist prognostication.

It is in his *Antiprognosticon* of 1560 and his *A goodly gallerye* of 1563 that Fulke most compellingly puts forth the case that natural phenomena, celestial and meteorological alike, could only remind one of God's judgment and mercy but should never be used for making specific prophesies. Printed eight times between 1563 and 1670,[27] Fulke's *A goodly gallerye* was popular, due to the favor Fulke found with Leicester as a patron as well as also to his clear, intuitively satisfying description of key meteorological processes derived in most respects quite precisely from Aristotle but with critiques of Aristotle by way of additional information derived from Seneca, in *Natural Questions* (c. 64 AD), and Pliny the Elder, in the *Natural History* (c. 77 AD). In addition, Fulke's organization is both complex, matching the subject matter, and rationally structured for reading, versus that which Aristotle conceived for delivery in a lecture and to serve a larger program of mapping the physical sciences. For example, Fulke first divides *meteora* between "imperfectly mixed," such as tempests and rainbows that are fleeting in constitution, and "perfectly mixed" that include gems and minerals. He then treated them by those *meteora* associated more with moist versus dry exhalations. Throughout the text, he added his own strategy to Aristotle's to make the meteorology more appealing to his readers.[28]

The popularity of *A goodly gallerye* was also due to Fulke's impressively engaging efforts to explain to those unfamiliar with this science just how it mattered to them—just how it related to the rest of their lives as members of a community, of England, and as faithful Christians each with his or her own reasoning capacity and informed conscience. We hear his effort reflected in the lengthy title of the work, which is worth noting in full: *A goodly gallerye with a most pleasaunt prospect, into the garden of naturall contemplation, to behold the naturall causes of all kynde of meteors, as wel fyery and ayery, as watry and earthly, of whiche sort be blasing sterres, shooting starres, flames in the ayre &c. tho[n] der, lightning, earthquakes, &c. rayne dewe, snowe, cloudes, springes &c. stones, metalles, earthes &c. to the glory of God, and the profit of his creaturs.* With this, his *goodly gallerye*, readers might, he adds, avoid the fate of "ignorant men [who] judge of these thynges that they knowe not," turning them into the work of the devil or the signs of purgatory (sig. B3r-v). Fulke wanted to show that instead, the works of God speak naturally rather than supernaturally of God's merciful plan rooted in identifiable materials and motions of the sublunary realm. One additional and crystalizing example is revealing, and for this I am indebted also to Crane's work, citing Crane in a return to Fulke's *Antiprognosticon*, in which he boldly takes on the events of the book of Genesis:

> Fulke offers an extended discussion of Genesis 1:14, insisting that the "signs" and "tokens" mentioned in the Bible are simply signs of natural phenomena themselves. He therefore directly articulates the view

> that nature offers only signs of its own workings: the signs mentioned in Genesis 1:14, are not "signes . . . for ye purpose of their predictions" but instead "they shalbe tokens of spring, somer, autume, and winter, of dais natural and artificial, long or short of years, according to ye sons course, or to ye moones course(B5v–B6r)."

These signs should be read not politically to forecast or justify changes in government but practically to allow merchants, farmers, and pharmacists to practice their needful crafts; hence, Fulke would argue, the need for his *A goodly gallerye*, to aid these practitioners and at once to shut down dangerous speculation regarding the meaning of things just like earthquake of 1580. By the time of the earthquake, key points of a materialist meteorology were common knowledge in England, and this is largely due to Fulke.

*

The Christian cosmological paradigm as I am tracing it here dates to the beginning of western Judaic practice that emerged in part as a response to both the polytheistic and materialist paradigms. It posits a human-aware creator God as the cause for all matter and motion in the universe, sharing with the polytheistic paradigm a sentient, passionate cause for creation; at the same time, it seeks to account for changes in sublunary sphere by way of causality based not on the passion of a being more powerful than humans but on that being having worked through nature to bring about creation, relying on primarily standardized and observable materials and motions. In the Christian paradigm, all *meteora* are thus materially constituted and operational but they are so composed and placed in motion to effect God's long-term plan for humans. All *meteora* in the Christian paradigm are both meteoric events and teleological things with God-assigned meaning: they are certainties not uncertainties; essential plot points, not accidents. Noah's Flood, the rainbow after it, the parting of the Red Sea, the plagues on Egypt, and the earthquake at Christ's crucifixion—all, and all almost equally, are in a Christian cosmology so many teleological stepping stones. All equally in the very end signify a new heaven and earth for those who are found righteous.

It was essential for many Christians, including St. Augustine but also many early modern Elizabethans, to believe that not only rare or easily noticeable *meteora*, but all meteorological phenomena down to the most obscure, were specially designed messages reinforcing a biblical and Christian interpretation of all events in the universe. Augustine explains this in *Confessions*:

> [I]n the separate parts of your creation there are some things which we think of as evil because they are at variance with other things. But there are other things again with which they are in accord, and

> then they are good. And all these things which are at variance with one another are in accord with the lower part of creation which we call the earth. The sky, which is cloudy and windy, suits the earth to which it belongs. . . . For all things give praise to the Lord on earth, monsters of the sea and all its depths; fire and hail, snow and mist, and the storm-wind that executes his decree; all of you mountains and hills, all fruit trees and cedars; all you wild beasts and cattle, creeping things and birds that fly in air; all you kings and peoples of the world.[29]

All creations, biotic and abiotic, signify beyond themselves as part of God's creation, greater than the sum of its parts but dependent on each part, "monsters," "hail," "mist," "trees," "people." All equally—"all you"—signify and contribute to the Christian story that will end in a new heaven and new earth; thus, Augustine concludes, "I no longer wished for a better world, because I was thinking of the whole of creation . . . The sum of all creation is better than the higher things alone" (149). In this system, no one need pay attention to the matters and motions of one of the *meteora*, because each and every configuration of the four elements points to the same origin and outcome. Best to limit consideration of wonders to these parameters, turning to emphasis on sin and virtual rather than on matter and motion.

St. Augustine also opened the door to the synthesis of polytheistic and Christian philosophies by way of Neoplatonism. Part of his conversion to church-accepted belief is coming from Manichean beliefs in a vital materialism, with God in all things, to beliefs inspired by Plotinus (204/5–70), a Platonist who held that God was not in matter and that evil is caused by forgetting that God instead is an intellectual Unity or Oneness separate from matter.[30] For these reasons, in this volume, I consider Neoplatonism as a version of the Christian model; sometimes it appears in more or less overt form in early modern textual representation of and explanations for meteors, most notably those following Giovanni Pico della Mirandola (1464–93). It does not, however, offer a viable alternative paradigm for the experience, knowledge, or representation of the meteorophysiological condition, and it is for this reason I do not consider it as a separate paradigm.

Writing in the thirteenth century and aiming similarly to offer a synthesis of pagan and Christian thinking about the created world, Thomas Aquinas took up Aristotle, helping to make room for the university study of Aristotle among theologians and within universities. His work would influence later humanists, such as Thomas More, Thomas Linacre, and others who in England helped to found the College of Physicians under Henry VIII. With respect to the interpretation of *meteora* and change generally, Aquinas saw as the same force the Christian God and Aristotle's prime mover, both of whom reside in the materials of creation and govern their motions from a distance by way of causation:

> The fifth way [to prove God's existence] is taken from the governance of the world. We see that things which lack intelligence, such as natural bodies, act for an end, and this is evident from their acting always, or nearly always, in the same way, so as to obtain the best result. Hence it is plain that not fortuitously, but designedly, do they achieve their end. Now whatever lacks intelligence cannot move towards an end, unless it be directed by some being endowed with knowledge and intelligence; as the arrow is shot to its mark by the archer. Therefore some intelligent being exists by whom all natural things are directed to their end; and this being we call God.[31]

Even those "things which lack intelligence" are "directed" toward "their end" by their material-dependent causation. They have individual ends, not only eschatological ones, but they also owe their individual beings to "some intelligent being . . . we call God." With this reasoning, *meteora* can have material and efficient cause as per Aristotle's observations, and they can also have final causes that, Aquinas seems to say, even Aristotle might have accepted, provided one substitute "God" for Aristotle's "prime mover." The result is the conclusion "that not fortuitously, but designedly do [all natural bodies] achieve their end." Although in the end, the *meteora* are nevertheless indistinguishably part of a larger plan designed by God, there is reason to examine their distinct qualities; the better to understand them in their detail is also to ponder and better know God's plan. It is to express faith.

This integration of materialist and Christian beliefs was easily and widely adopted. Among the most memorable and influential of its applications occurs in Dante's *Divine Comedy* (c. 1320). Its massive, heavy Satan dwells in a cold Hell at the dead (and we might say doubly dead) center of the geocosmic universe, within earth, in its center to which all damned souls. Angels and beatified souls, being light and aflame with religious love, naturally instead rise through the fiery regions of the upper atmosphere and beyond to reside with God in the heavens. They reside at the margins of the universe from the fallen human perspective, but the circumference is the center from a Christian, God-centered view. Sublunary meteorophysiological phenomena in the *Commedia*—its earthquakes, boulders, thunder, and represented bodies of suffering and beatified souls—exist for God's purpose to advance or challenge humans on their individual pilgrimages, but these phenomena are individually construed indicating Aristotelian more than Platonic influence: memorable, imaginative, and unique in their details, they are materially authentic even as they are part of God's salvific plan.

By the reign of Elizabeth I, the three cosmological paradigms served well and widely, pressed into service in blended forms across confessions, genres, classes, and disciplines. Adherence to a Christian interpretation of all wonder was politically advantageous and generally safe, but there were factions among Christians who could not agree on the origin, effects, or meaning of that 1580 quake, let alone on things like monstrous births and strange murders. These factions also employed the three paradigms as they responded to each other's claims. The rational and experientially-validated appeal of the materialist and thrill of the polytheistic explanations for creation and *meteora* gave writers room to argue, to differ, and to move readers in delight that might also be instructive and memorable. The flexibility for representation of lived experience that was for this reason afforded writers in a post-Reformation and pre-scientific era is remarkable.

The 1580 earthquake pamphlets that are the focus of investigation in the next chapter provide an entry into this flexible, complex, wondrous space of early modern experience. This space is what Kuriyama calls a "Once upon a time" in which "human being was being embedded in a world." In entering this space and exploring it, there may be monsters but there will also be opportunities to think outside of the boxes of our own cosmological paradigm, which is decided less accessible or flexible than the early modern trio that this volume examines. An approach to early modern meteorophysiology in this context, then, specifically grants practice in the cultivation of an open-mindedness—in Bennett's words, of a "naiveté" by way of "revisit[ing] and becom[ing] temporarily infected by discredited philosophies of nature" (18). And with this open mindedness, we might learn, or remember, that one person's giant is another's earthquake, one's earthquake is another's angry God.

Notes

1. Among those ahead of the curve are Vladimir Janković, *Reading the Skies* (Chicago: University of Chicago Press, 2000), and Craig Martin, *Renaissance Meteorology: Pomponazzi to Descartes* (Baltimore: Johns Hopkins University Press, 2011).
2. Katherine Cox, "The Power of the Air in Milton's Epic Poetry," *SEL: Studies in English Literature, 1500–1900* 56.1 (2016): 149–170. Cox offers important insight on Milton's exploration of the meteorological phenomenon of the exhalation and its relationship to orality in Paradise, as the "mists and exhalations" are "voicing God's praise" and as "the earth vocalizes her felicity with 'fresh gales and gentle airs'."
3. Oxford English Dictionary, s.v. meteor, noun, A1 and 2. For more on this word's meanings, see Janković, *Reading the Skies*, 15–16.
4. The choice of "polytheistic" over "pagan," "classical," and other options is with an aim both for accuracy and ease of comprehension and use. This term is ahistorical and problematic, but to call this cosmology by any other name poses more problems for one or the other consideration. The same is the case for the choice of "materialist" over the options of "secular" or "scientific" to describe the paradigm most associated with Aristotle. For more on the terms, see each in Edward Craig, *Routledge Encyclopedia of Philosophy* (London: Routledge, 1998). Here Christian means Judeo-Christian, with

first codification in the Hebrew Bible. It is also the case that the polytheistic paradigm predates its codifier Hesiod, as Todd Borlik reminds us, usefully in his treatment of Pythagoras who, he says, is credited with coining the Greek word "kosmos" (*Ecocriticism and Early Modern English Literature: Green Pastures* [New York: Routledge, 2010], 58. 34). The modern view of polytheism is far more fully and directly informed by Ovid, so I use Hesiod's codification here for simplified accuracy, but a closer examination of the Renaissance Ovid, Pythagoras, and Lucretius by way of meteorophysiology and in conversation with Neoplatonism would be of value, taking up both Borlik's findings and those of Gerrard Passannante (*The Lucretian Renaissance: Philology and the Afterlife of Tradition* [Chicago: The University of Chicago Press, 2011]). In this volume, these are threads smaller than others, impossible also to take up alongside more prominent ones.

5. Hesiod, *The Theogony*, Hugh G. Evelyn-White, translator (New York: Loeb Classical Library, 1914), 11.116–138. See Evelyn-White's notes, including this explanation of "earth" and "deathless ones": "Earth, in the cosmology of Hesiod, is a disk surrounded by the river Oceanus and floating upon a waste of waters. It is called the foundation of all (the qualification 'the deathless ones,' etc. is an interpolation), because not only trees, men, and animals, but even the hills and seas are supported by it" (11.129, 131).
6. For more on the Lucretian strand in Ovid with relationship to later discussions of Arthur Golding's translation and Gabriel Harvey's send-up of those efforts, see Gerard Passannante, "The Art of Reading Earthquakes: On Harvey's Wit, Ramus's Method, and the Renaissance of Lucretius," *Renaissance Quarterly* 61.3 (2008): 792–832.
7. As a starting place for examination of Ovid's legacy, see Alison Keith and Stephen J. Rupp, *Metamorphosis: The Changing Face of Ovid in Medieval and Early Modern Europe* (Toronto: Centre for Reformation and Renaissance Studies, 2007).
8. Todd A. Borlik, *Ecocriticism and Early Modern English Literature: Green Pastures* (New York: Routledge, 2010), 58.
9. Lynn Enterline, *The Rhetoric of the Body From Ovid to Shakespeare* (Cambridge: Cambridge University Press, 2000), 12; Liz Oakley-Brown, *Ovid and the Cultural Politics of Translation in Early Modern England*, Studies in European Cultural Transition (New York: Routledge, 2006), 12. For more on the power of Ovid over early modern imaginations, see Jonathan Bate, *Shakespeare and Ovid* (Oxford: Clarendon Press, 1994); A.B. Taylor, *Shakespeare's Ovid: The Metamorphoses in the Plays and Poems* (Cambridge: Cambridge University Press, 2000); Cora Fox, *Ovid and the Politics of Emotion in Elizabethan England* (New York: Palgrave Macmillan, 2009); and Raphael Lyne, *Ovid's Changing Worlds: English Metamorphoses, 1567–1632* (Oxford: Oxford University Press, 2001).
10. Ovid, *Metamorpheses I*, Frank J. Miller, translator, Loeb Classical Library (Cambridge, MA: Harvard University Press, 1977), 9 (cf. Latin lines 78–84).
11. Arthur Golding, *The XV Bookes of P. Ovidius Naso, Entytuled Metamorphosis*. Translated out of Latin into English meeter by Arthur Golding Gentleman, *A Worke Very Pleasaunt and Delectable*, 1567, sig. B2v-r.
12. Aristotle, Physics, H.D.P. Lee, translator, Loeb Classical Library (Cambridge, MA: Harvard University Press, 1952), 1:555, 26.
13. See as a popular starting place, Peter Tyson, "The Star In You," *NOVA Science NOW*, posted 12.02.10 (www.pbs.org/wgbh/nova/space/star-in-you.html, accessed 3 June 2017); and Muhammad Nafees, "An Inquiry Into the Cosmic Origins of Human Life," *Medical Journal of Islamic World Academy of Sciences* 18.2 (2010): 85–90.

14. For more on the relevance of this introduction with respect to Aristotle's vision not only for his studies into and codification of knowledge related to natural phenomena, see Andrea Falcon, *Aristotle and the Science of Nature: Unity Without Uniformity* (Cambridge: Cambridge University Press, 2005), especially chapter 1. In most cases, however, scholars cite this introduction to *Meteorology* only to move quickly to Aristotle's treatment of plants and animals.
15. Malcolm Wilson's *Structure and Method in Aristotle's 'Meteorologica': A More Disorderly Nature* is which is essential reading on this subject (Cambridge and New York: Cambridge University Press, 2013), 281.
16. Wilson also calls the exhalations "dualizing" entities, which Aristotle saw as able to cross boundaries among all sublunary species, meteoric and terrestrial alike (see especially 113–114).
17. Martin, *Renaissance Meteorology*, 4.
18. Etienne Gilson, *From Aristotle to Darwin and Back Again: A Journey in Final Causality, Species, and Evolution* (Notre Dame, IN: University of Notre Dame Press, 1984), 2–3.
19. As Martin explains, "If meteorology considers objects that are without their own substantial forms, knowledge of formal causes would be limited" (27). Aristotle would choose not to entertain for the *meteora* these causes, instead concluding that it was fruitless to even make the effort. He left the *meteora* as lacking in formal or final causes. On the matter of Aristotelian causality and the *Meteorology*, see Wilson pages 93–97—this his section called "Final Causes."
20. Nearly all early modernist natural philosophers associate earthquakes and thunder by way of the processes they have in common: air trapped in earth or in a cloud/water, each then exploding forth. For a summary of the earthquake and thunderclap relationship see Janković, *Reading the Skies*, 18, and for a longer treatment of the ways that certain *meteora* and other physiological phenomena resemble each other, see his chapters 2 and 3 on the comet, sigh, earthquake, and plague in particular.
21. William Shakespeare, "*Macbeth*," in Sandra Clark and Pamela Mason, editors, *The Arden Shakespeare*, Third Series (London; New Delhi; New York and Sydney: Bloomsbury Arden Shakespeare, 2015), 5.5.26–27.
22. Michael Witmore, *Culture of Accidents: Unexpected Knowledges in Early Modern England* (Stanford, CA: Stanford University Press, 2001), 2, 19.
23. William Fulke, *Antiprognosticon that is to saye, an invective agaynst the vayne and unprofitable predictions of the astrologians as Nostrodame, [et]c. Translated out of Latine into Englishe. Wherunto is added by the author a shorte treatise in Englyshe, as well for the utter subversion of that fained arte, as also for the better understandynge of the common people, unto whom the fyrst labour seemeth not sufficient* [Imprinted at London: By Henry Sutton dwellyng in Pater noster row at the signe of the blacke boy, the. 23. day of December. And are there to be solde. Perused and allowed according to the Quenes majesties injunctions] (1560); *A goodly gallerye with a most pleasaunt prospect, into the garden of naturall contemplation, to behold the naturall causes of all kynde of meteors, as wel fyery and ayery, as watry and earthly, of whiche sort be blasing sterres, shooting starres, flames in the ayre &c. tho[n]der, lightning, earthquakes, &c. rayne dewe, snowe, cloudes, springes &c. stones, metalles, earthes &c. to the glory of God, and the profit of his creaturs* (1563).
24. Stuart Clark, *Thinking With Demons: The Idea of Witchcraft in Early Modern Europe* (Oxford, UK: Clarendon Press, 1997), 11, 534–535. Please see his much more developed description of the Protestant efforts to redefine the

supernatural, wresting control for its definition from Protestans. Please also see Mary Thomas Crane, *Losing Touch With Nature: Literature and the New Science in Sixteenth-Century England* (Baltimore: Johns Hopkins University Press, 2014), 15; Kristen Poole, *Supernatural Environments in Shakespeare's England: Spaces of Demonism, Divinity, and Drama* (Cambridge: Cambridge University Press, 2011).

25. For more on these men's writings related to meteorology, and on the physical world generally, see Crane, *Losing Touch With Nature*, 68–75; Stephen Johnston, "Digges, Leonard (c.1515—c.1559)," *Oxford Dictionary of National Biography* (Oxford: Oxford University Press, 2004; www.oxforddnb.com/view/article/7637, accessed 24 July 2017); R. Julian Roberts, "Dee, John (1527–1609)," *Oxford Dictionary of National Biography* (Oxford: Oxford University Press, 2004; online edn., May 2006, www.oxforddnb.com/view/article/7418, accessed 24 July 2017).
26. For more on Fulke, see Richard Bauckham, "Fulke, William (1536/7–1589)," *Oxford Dictionary of National Biography* (Oxford: Oxford University Press, 2004); Janković, *Reading the Skies*, 24–25; and Richard Bauckman, "Science and Religion in the Writings of Dr. William Fulke," *British Journal for the History of Science* 8.1 (1975): 17–31. Thanks to Crane again for calling my attention specifically to Bauckman's stronger claims regarding Fulke, including his assertion that Fulke's *goodly gallerye* is "the only properly scientific discussion of the subject in English in the sixteenth century" (28, cited by Crane, *Losing Touch With Nature*, 72).
27. To this point of dating a "general knowledge" of "Aristotle's explanation of earthquakes" to "the end of the sixteenth century," see R.M.W. Musson, "A History of British Seismology," *Bulletin of Earthquake Engineering* 11.3 (2013): 727. See also Janković, *Reading the Skies*, 23–25.
28. In the case of distinguishing between imperfectly and perfectly mixed *meteora*, Fulke drew from Aristotle's *On the Heavens* [448 2.269a.1–15], and in the case of exhalations, he used the definitions of Aristotle but he supplied a clearer structure in their presentation—thus in both cases remaining true to the original while suiting it to his reader's needs or at least to his desire regarding their use of it.
29. St. Augustine, *Confessions* (New York: Penguin, 1961), 7.13 pages 148–149.
30. A short primer on Neoplatonism follows in this note, the better to demonstrate that with respect to its representational possibilities that Neoplatonism does not stand distinctly enough from Christianity to find it as a distinct through line in most early modern texts, in contrast to pagan/Hesiodic, materialist/Aristotelian, and Christian paradigms which are distinct and competing representational registers. Like Platonism itself, which does not either posit a distinct enough representational register for meteorophysiology, Neoplatonism posits a masculine force that, with the help of a female world soul, imprints female matter, giving it form—a vestige of the polytheistic model being the female material being separate from God. As Carolyn Merchant explains, in this paradigm, a female soul acts as "the bridge between the unchanging eternal forms and the changing, sensible, temporal lower world of nature" *The Death of Nature: Women, Ecology and the Scientific Revolution* (New York: Harper and Row, 1983), 10. As Mary Garrard has recently reminded us, however, Plato's concept of a demiurge was part of a general move toward the privileging of male-dominant creation narratives over female-dominant ones—a move embraced in particular by the Medicis who were instrumental in the funding of the Renaissance, *Brunelleschi's Egg: Nature, Art, and Gender in Renaissance Italy* (Berkeley and Los Angeles: The University of California Press, 2010), 9–30. In Platonism and Neoplatonism,

the male becomes the artist/architect, the female the material he shapes—a reading like that which Arthur Golding supplies in his interpretation of Ovid. The demiurge as artist might then be seen both to precede and to create both the female the world soul and the female matter, including earth; as Plato explains in Timaeus: "The earth, which is our nurse, clinging around the pole which is extended through the universe, he framed to be the guardian and artificer of night and day, first and eldest of gods that are in the interior of heaven" (Plato, "Timaeus," in Edith Hamilton, Huntington Cairns, and Lane Cooper, editors, *The Collected Dialogues of Plato, Including the Letters* (New York: Bollingen Foundation, 1963), 1169. This creation story for earth stands distinctly from that of a strictly Hesiodic earth who comes into being without explanation and possibly of her accord and who then births Heaven/Sky with whom eventually she also creates all other tangible, beings. As mentioned, it so closely resembles, however, many early modern Christian accounts of the relationship between nature and God. With respect to the subjects of meteorology and physiology, it flattens them out as Christianity does, rendering them unfit for investigation in their own right, each material thing a distraction from the Forms beyond them.

31. St. Thomas Aquinas, *The "Summa Theologica" of St. Thomas Aquinas: Part I, Qq. I-Xxvi* (London: Burns, Oates & Washburne, 2008), online, question 2 article 3.

2 The Sneezing of the Earth

When it rocked parts of England and the Continent on 6 April 1580, the Dover Straits earthquake called attention directly to the material vulnerability of all bodies and the earth itself. An extraordinarily rare event for England, this quake struck in the evening, around six o'clock, lasted about 60 seconds, and by current estimates measured at between 5.3 and 6.0 on the Richter scale.[1] From here, we lose sight of current seismological statistics and move into the phenomenology of the sixteenth and seventeenth century earthquake experience—what seismologist R.M.W. Musson calls "observational seismology." In premodern history, "earthquake monitoring was restricted for the most part to documenting earthquakes through their felt effects, not always scientifically" (716). The first published pamphlets on the subject record exactly these effects, with repeated details of bells tolling on their own, people fleeing from buildings in fear for their lives, others in theater balconies leaping to safety, and an apprentice and servant dying as a result of a stones falling from a church rooftop.[2]

The quake shook people into an inquisitiveness that was expressed in many forms, personal and public, including two dozen entries in the Stationer's Register in the days and weeks following it. There had been other earthquakes in Britain prior to 1580, but none gained as much attention—no doubt because of the printing press, helping establish the determination that this had been a "general earthquake" felt throughout Britain. Although the material damage done by the quake was slight given the areas it spanned, quakes are rare in England, and this was the largest not only in memory but as far as any of the Chronicle entries had recorded. Musson confirms as well regarding the prior century: "The record of seismicity in Britain in the fifteenth century is almost a complete blank" (2013, 719). It is little surprise that some declared the 1580 quake to be universal in impact, and Todd Borlik is correct, and delightfully so, to explain it this way: "the 1580 earthquake seems to have rattled the Elizabethan World Picture."[3] Historians added it to Chronicles; Shakespeare likely had it in mind as a way to enhance the character of Juliet's nurse (to be discussed in chapter 5); and individuals including Thomas Trevelyon added it to their lists of significant world events (fol. 3). Trevelyon's private, unpublished miscellany includes as one of its first pages "A brief computation of this time complete within this present year: 1608." It begins its count of 3,914 years on its first page "From Noah's flood unto this present year." Among the 32 events deemed worthy of record, punctuating all "time complete" and beginning with the "Creation of the world" itself, Trevelyon offers as the most recent dated event on the list, "The general Earthquake in England." This was 28 years after the 1580 quake, and it still loomed large in imaginations and memories; perhaps it had grown there, sewn with alacrity into personal and national biographies.

For many, the earthquake imprinted the body-mind and altered all other sensory-related terrain immediately, making its popularity as a subject for pamphlets easy to anticipate. The quake demanded interpretation and, as it was a subject of epic proportion and of immediate consequence, it would be certain to turn a profit for printers. The first entry in the Stationer's Register appeared the very day after the quake. It was a ballad entitled *a godly newe ballat moving us to repent by the example of the e[a] rthquake happened in London the.* 6 *of Aprill* 1580. The ballad does not survive, but four full entries with "earthquake" in the title appear the very next day, and a reading of several of the surviving earthquake pamphlets entered in the register between April 8 and June 30 illustrates clearly the struggle in the period to account for differences between natural and supernatural events generally and to place properly in understanding this earthquake and the complex responses it generated just as suddenly as and perhaps more disturbingly than the literal phenomenon. In this chapter, I examine in order two inexpensive, popular pamphlet responses to the earthquake, each one notably different; the Church of England's schedule of prayers created in part to answer some of those pamphlet responses; and an hilarious, satirical send-up of all prior earthquake responses. The set as a whole illustrates the remarkable range of representational registers available for the expression of the earthquake experience in this period.

*

Among the writers responding to the earthquake was lifelong soldier and writer, Thomas Churchyard (1523?—1604). His is the earliest of the earthquake pamphlets that survives for examination, and in it, Churchyard primarily cries out against sin and calls on readers to examine their consciences; these, he says, are the only responses to the earthquake experience. All others may lead to hell. Little else concerns him, it seems. Yet, it is as immediately apparent that his own social standing matters as much as his major claims regarding the quake. The full title of his pamphlet contains some initial signs of these efforts: *A warning for the wise, a feare to the fond, a bridle to the lewde, and a glasse to the good. Written of the late earthquake chanced in London and other places, the. 6. of April 1580 for the glorie of God, and benefite of men that warely can walke, and wisely can judge. Set forth in verse and prose, by Thomas Churchyard Gentleman. Seen and allowed.* The pamphlet is indeed a mix of "verse and prose" that includes a ballad, prayers, and three sections that in one way or another are a "report" of the earthquake. The pamphlet's prefatory sections demonstrate still further what we might describe as notably overt efforts to secure his reputation as an authority on this subject. His unusually bold dedication to the well-established, Calvinist Dean of St. Paul's, Alexander Nowell (1516/17–1602) is a case in point. He proclaims to Nowell, "I have chosen you among the multitude to sounde out the trumpet of Gods glorie" exemplified by the recent "wonders of the Lorde" in the earthquake (sig. A1v-A2r).[4] Churchyard leads with an effort to make his message, himself, and the quake memorable and serviceable.[5]

In the second prefatory section, self-attentively called "Churchyardes admonition to the Reader," he extends his exhortation by an appeal to the role that memory plays in the identification and interpretation of wonders:

> Trust (good Reader) that the fresh memorie of this late wonder shall make thee to look back-warde unto thy former faultes, and make thee a new man in cleanness of life. For the stranger the things are that our eyes beyond, the more the impression of the mind is earnestly occupied about the understanding of a wonder.
>
> (sig. A2v)

Churchyard's theory of memory—that a great wonder makes more of an imprint upon the "the mind"—originates in classical sources (such as Marcus Tullius Cicero's *Rhetorica ad Herennium*), and it is in keeping generally with early modern thinking about and experience with memory. The quake will be remembered, but because it is the natural inclination of the "mind" to "earnestly occup[y]" itself in response to anything that does not easily fit within prior experience, Churchyard reveals quickly and explicitly the danger of such earnest occupation. His readers, he imagines, have already likely turned to sources other than their consciences or the Bible for the interpretation of the quake. To address those busy minds, he acknowledges:

> [P]erhaps, some fine headed fellowes will wrest (by naturall argumentes) Gods doing and works, to a worldly or earthly operation· proceeding from a hidden cause in the body and bowels of the earth. As in deede of many other earthquakes before, hath bene written and th[o]roughly disputed: A reason that man maketh, and a matter to be spoken of, but not much to be commended. Let such fine wittes search out secretes, and sift what they can from the bottome of their senses. Yet those that feare God, (and feel in their consciences a divine motion from the consideration of worldly wonders) will take

> the Earthquake to be of a nother kinde of Nature: And beholding the myraculous manner of the same, with open armes, and humble heart, will embrace Gods visitation, & worthily welcome the messenger he sendeth.
>
> (sig. A2v–3r)

Churchyard contrasts the foolish "fine headed fellowes," espousing materialist reasoning to account for the earthquake as a phenomenon of nature, with the wise Christians who turn inward instead to listen to "their consciences." There Christians will find a God-sent "visitation" of "divine motion" mimicking the earthquake and calling them to repent. For Churchyard, fear and trembling are valuable instigators in the incitement to faith but they can lead one to fall prey to the seductive "naturall arguments" that are "a reason that man maketh, and a matter to be spoken of, but not much to be commended." By this reasoning, Churchyard intends his message to be an "admonition," almost a fatherly message of warning not to put oneself in league with such "fine wittes" that only "sift what they can from the bottome of their senses."

This passage tells us what Churchyard does not in so many words: these "naturall arguments" about "a hidden cause in the body and bowels of the earth" pose a serious threat to faith such that his warnings and other efforts are necessary to contain them. It is as if natural arguments were some secret and dangerously attractive formula, like the words of the witch that Stavreva calls "witch-speak."[6] These "arguments" are based on the idea of "a hidden cause," he says, refusing perhaps to give more specific detail from materialist earthquake theory regarding wind trapped in subterranean hollows. It is not overtly risking overstatement to suggest that for some Christians, the threat posed by materialist meteorology was clear and present with the earthquake of 1580. This may be a reason for Churchyard's dedication of his pamphlet to Nowell. The Dean himself would soon be in the process of helping to devise a Church of England sponsored interpretation of the earthquake, as he had done during plague visitations of the past.

By the end of the admonition, Churchyard returns to the power of memory as part of his closing appeal to "the well disposed person" who "will be the better, while he liveth, as oft as this late Earthquake shall come to his minde and memorie" (sig. A4r), there to shake again the individual with fear and trembling as well as gratitude for a just and merciful God. Churchyard follows the Christian prescript for attempting to transform destabilizing instances of radical material change into events remembered in a context of faithful anticipation of a second coming, a resurrected body, and a new heaven and earth. This is a move that we will see John Donne make with much greater complexity and import, as his earthquake and each moment of destruction wrought on his bedridden body he will transform into a spiritual pulse that is also a prayer and a clarion call to faith—all necessary to ward off the doubt and despair more viscerally associated with his condition.

Churchyard also offers prayer in his pamphlet, but it is of a very different nature than Donne's. After describing the birth and death of Christ, the speaker petitions God, reinforcing the through line of memory:

> For remembraunce of these, and all thy other graces when thy sword of wrath is lifted up ready to confound us which rebel in sin against thy glory, then remember thy sonnes death, cast aside thy pitieful eyes, beholde his greevous wounds that bleed a fresh to pacifie thy anger, forget (O Lord) the offences of our youth, blot out of remembraunce the sinnes of our forfathers, & as it hath pleased thee to send so gentle a remembrance as this earthquake when it was in thy hands to shake our bodyes to dust, and our soules to hell.
>
> (sig. D1r)

The passage, much like the language of plague sermons in the period, addresses God to remind him that although his anger and punishments are justified, his mercy is all the more necessary for humans whenever he is so angered; further, that mercy, is already actively engaged, because every time his people sin, Christ's wounds "bleed a fresh," and because he (God) could be conducting a much more destructive "shak[ing]" of "our bodyes to dust, and our soules to hell." This logic aligns with the oft made statement by Luther that God could at any time bring together the waters of the heavens and of the earth in a second flood but that he always, daily, holds them apart as part a constant show of his mercy. This too will be the logic present in the Church of England's special prayers against the quake that circulate later in the same month.

Rather than conclude with this prayer, however, Churchyard provides verse reinforcement, the better perhaps to entertain along the way and establish a readership. The poem he shares is not even his own but comes, surprisingly, attributed to Richard Tarleton, the clowning master of human spectacle. Abraham Fleming lists Tarleton as among those writing on the quake, independent of Churchyard, but no separate publication exists. In quite a change of pace, then, Churchyard advances his argument regarding meteorophysiological wonder and memory:

> When Mountaines move as late they did in wales
> great signe it is it nature then is crost:
> . . .
> When blasing starres, and bloody cloudes doe show
> then time it is for men too search anew:
> And mark it stock from whence these grafts doe grow
> the fruit more straunge then any gardner knew
> the Aire is chokt with vapour of our sin:

When such unwoonted tokens call us in.

. . .

But if these tokens which be past and gon,
have took no roote at all within your harts:
You needs must graunt this earthquake to be one,
unlesse you chalenge Heaven for desarts.
Our health of soules must hang in great suspence
When earth and Sea doo quake for our offence.
(sig. D2r-3r)

The speaker "tell[s]" a "tale" that he promises by its wondrous content will compare with any marvel heard from abroad. So powerful will it be over imagination and memory, it is "No toy, no trifle, nor surmised jest:/ But worthy wel to lodge in every brest." It will be lodged in memory that is bodily. By these methods, Churchyard continues to draw his through line linking the disparate texts of his pamphlet, all reinforcing a single message: "this earthquake" is a "token" sent by God—a call to remembrance, a flashback in a full-body re-performance of the fear and trembling of faith. It should "roote . . . within your harts" as indeed human sin has caused "earth and Sea [to] quake for our offence." There, in "hart" it should fix, ever a reminder of what—in spite of Churchyard's varied delivery techniques—is rather relentlessly a single conclusion on the meaning of the quake.

*

Of course, people did not stop pondering the cause and import of the 1580 quake. The desire, even need, to contemplate it was itself an attendant effect of the quake, a phenomenon experienced broadly in Britain. On April 11, still only days after the quake had struck and slightly fewer than that to Churchyard's pamphlet, physician Thomas Twyne offered his more nuanced, and hesitant, treatment of the subject. By title, the pamphlet is in many ways similar to Churchyard's: *A shorte and pithie discourse, concerning the engendring, tokens, and effects of all earthquakes in generall: particularly applyed and conferred with that most strange and terrible worke of the Lord in shaking the earth, not only within the citie of London, but also in most partes of all Englande: Which hapned upon Wensday in Easter weeke last past, which was the sixt day of April, almost at sixe a clocke in the evening, in the yeare of our Lord God. 1580. Written by T.T. the 13 of April. 1580*. As Churchyard does, Twyne asserts that God is the cause of the quake. Where Twyne differs enormously from Churchyard and sets pace for writers who follow is in his extensive treatment of the very natural arguments regarding earthquakes that Churchyard had proclaimed dangerous.

By 1580, Twyne was well known, in a good position to enter with some security into this arena, already having translated the works of other "prophane wryters" (sig. A1r). He had completed a translation of *The Aeneid*, begun by Thomas Phayer.[7] He also translated a number of treatises with a materialist leaning, including Petrarch's *Physic Against Fortune* (1579), Petrarch's guide against all terrifying life challenges including earthquakes, tempests, and plague.[8] The surprise might be that his earthquake pamphlet was an original piece of writing, and it is fascinating the degree to which he explains at the start why it is that he relies so heavily on scholarly learning to account for the quake:

> 2. But before we enter any further into the bare bewraying of the matter, it is expedient that I discover unto you the causes, and substaunce of everie Earthquake, which I must be fayne to borrowe from the Prophane wryters, who have most dilligently laboured in the search of naturall causes, wherunto doubtlesse they could not so clearly have atteyned without the finger of God, which hath led men as well into the true contemplation of these matters, as of any other knowledge. And therfore following *Aristotle* as chief in this behalfe: we must understand, that the efficient causes of an Earthquake are three, to wyt, the Sun, the other sixe Planets, and a spirite or breath included within the bowelles of the earth: and the materiall cause one, which is an Exhalation, that is to say, a certaine ayre, breath, or smoake drawne out of the earth, which of nature is hot and drie.
>
> 3. Thus it is not hard then, to describe the engendring of an Earthquake.
>
> (sig. A1r-v)

How the earthquake is "engendr[ed]" is plain if one turns to Aristotle. Thus, by page one, Twyne has exposed Churchyard's fear: the materialist account can render all others unnecessary. But Twyne makes two moves to appropriate his Aristotle: first he suggests there is no other place to turn, that he "must be fayne to borrowe from the Prophane wryters" on this subject; and then he clarifies that Aristotle could not have known what he did about "naturall causes" were he not "led" by "the finger of God." Also in Twyne's terminology are the vestiges of a pantheistic respiring, generative (capable of "engendering") earth.

What Twyne aims to accomplish by his overt use of materialist cosmology he confirms several pages later, when he offers to "compare[s] some parte of these generall tokens and Accidents, with this our particular Earthquake" (sig. A3v). By his method, Twyne explains, "shall we bee the better able to discerne of this wonderfull worke of God, whether it be meere naturall, or no." It is at this point that he also admits that "Specifically on the former claim," he must:

> take the better occasion to report of every poynt thereof, according as I have beene enformed by persons of credite. For why? for mine

> owne parte, I must thus protest before the lyving God, whose matter we have in hande: that beeing not much past a payre of Butte lengthes without the libertie barres of the Citie of *London*, walking with honest godlye companye, and to my lykyng, even at the instant of the quaking, as it shoulde seeme, neyther they, nor I perceyved any such thing at all. But the Lorde hath his providence, and his workes are marveylous.
>
> (sig. A3v)

Twyne did not himself experience the shaking, and his sharing of it, couched in a careful context for his need to disclose the fact, is pronounced and suggests that he is aware of the ramifications of the claim. Others had claimed that the quake was universal in scope, and here, Twyne's testimony suggests that at minimum, if the claim was universal, it was so in ways not physically apparent. More pointedly, his wording makes explicit how earnest he is in the endeavor: "I must thus protest before the lyving God, whose matter we have in hande . . . even at the instant of the quaking, as it shoulde seeme, neyther they, nor I perceyved any such thing at all." He claims for himself that he is a man of "the lyving God" and that he understands that it is God's "matter we have in hande," which would be the original cause for the earthquake to God's purpose, but that just when other accounts pinpoint the time of the quake, "at the instant," not only did he not feel it himself but no one in his company "percyved any such thing at all." No one felt a thing, which may be why he adds "as it shoulde seeme." This is a strong, complex, socially-aware report of his own lack of personal experience with the quake.

British seismologist R.M.W. Musson uses Twyne's lack of feeling it himself, combined with Twyne's effort to track the "tim[ing]" of "the arrival of the earthquake in each place" (725) based on the claims from those who did feel it to track it, to claim Twyne's pamphlet is the one that most closely approximates a scientific treatise; it is as if Twyne attempts to take the personal out of what is a search for an epicenter.[9] Furthering this project, Twyne offers in the bulk of the pamphlet a point-by-point comparison of the natural features of an earthquake as they do and do not line up with the events of the 1580 quake—its accompanying sights, sounds, and temperatures, its kinds of motion and effects on water—all to conclude that the quake was indeed supernatural, not natural. But the thoroughgoing Twyne does change tack for a bit before proceeding in his examination. He waxes literary in ways that underscore the representational flexibility afforded these writers in an era prior to disciplinary distinctions:

> Indeed the suddainesse and strangenesse of the thing was such, that it tooke divers men in divers actions, and brought them into sundrie considerations of the matter. Some doubtlesse at their prayers, and

> hearing godly Sermons, whome, as men, it must needes amaze, or bring into a muse. Some at the Taverne, and upon their Alebench, and therefore might well suspect that it was long of their liquour. Some in earnest conference of worldly affaires, and so peradventure they tooke small or no regard at all of it. Some in ydlenesse alone, and those of likelyhoode it might sorely abash. Some at game, and therefore not muche moved. Some at common Playes, who as I understand, were horribly troubled. Some . . . imputed the ratling of wainescots to Rattes and Weesels: the shaking of the beddes, tables, and stooles, to Dogges: the quaking of their walles to their neyghbours rushing on the t'other side.
>
> (sig. B2v–B3r)

With an anaphoric "Some," Twyne effectively speaks of Londoners in all of their various pursuits—everyone, it would seem, reacting to the shaking (except himself, apparently). His comments also indicate an effort to establish the broadest of audiences for his pamphlet—all of the many " [s]ome"s—and to show that the quake is a phenomenon of epic proportion and import.[10] Twyne also captures the degree to which the quake calls people together in response, ironically, to the variety of their thoughts on the matter:

> And as their opinions were sundrie, so were their speeches therupon diverse, untill a common conference beeing had, they were resolved upon their common case & danger. For many not trusting to their own judgement, and partly also moved with feare, ran out into the streetes to know if the like had hapned unto others.
>
> (sig. B2v–B3r)

"Some" react this way, others that, but all are shaken into diverse opinions, on the one hand, and into collective certainty about the need to assign to the event some agreed upon significance, on the other. Worth additional attention in this portion of the passage is the line about people lacking in the trust of their own judgment, such a rare phenomenon this is. They come out into the streets to verify basic facts: did you feel it, how long did it last, what did you see and hear, are we safe, is anyone hurt, should we be afraid of more to come, what can we do about it, what can we say about it?

In answer, Twyne lacks Churchyard's zeal., and he seems aware of this lack as he his reader able to sense this and therefore demanding more in the way of conviction:

> Now, perhaps some would expect at my handes, that I should set down my judgement farther concerning the efficient causes, & also the consequents of this Earthquake by the position of the Heavens

> and aspectes of the Plancts, and fixed Starres, for that presente time: which now I must needes omit for brevitie sake, till some other time more convenient. And if likewise I were farther demaunded, what mine opinion is concerninge this Earthquake, whither I thinke it altogeather naturall, or not? Surely, I am otherwise perswaded, and so I judge many other to be, that have entred into the deep consideration therof.
>
> (sig. B4v)

Twyne anticipates readers wanting to hear from the almanac writer a thing or two related to the astrological significance of the quake, but for "brevity sake" Twyne defers. His answer on the natural versus supernatural cause of the quake is almost equally diversionary. Twyne comes close to saying that he believes the earthquake was supernaturally caused, but he does not, quite, instead opting for a statement that is articulated in the negative: "I am otherwise perswaded." He then moves away from himself as authority and to others, and even then he hedges: "so I judge many other to be." Continuing, he positions himself at another remove from the prior information he has shared to brush aside what anyone might deduce from "the deep consideration therof" and offer the following: "But let it bee, as it is, surely it cannot be without the speciall finger of God, whither it be for our comforte, or terrour, as every mans conscience shal beare him recorde, although I am sure there be none that can excuse them selves of sinne" (sig. B4r). We have even at this point an odd construction in "surely it cannot be," and again Twyne takes another step to separate himself as the authority from his reader's conclusions on the subject. The decision regarding whether this earthquake is natural or supernatural is not about learning, apparently, or about opinions developed through deep considerations. The conclusion can only be found "as every mans conscience shal beare him recorde." There, Twyne does suggest, one might look first to see not what the holy spirit is telling him or her directly about the earthquake but whether he or she has sinned. What is implied is that the sin alone tells the reader whether the earthquake was naturally or supernaturally caused.

This is Twyne's authoritative answer on the subject of the 1580 earthquake—such a contrast with Churchyard's pamphlet that its existence is evidence for Twyne's own description of people with "their opinions . . . sundrie" and "their speeches therupon diverse." On another score worth mentioning in brief is the contrast between the two writers with respect to national politics. Churchyard does not comment on the times in any quotidian fashion; his only mention of the queen, for example, is through Tarleton's poetry, with its recommendation to "first honour God, and then obey your Prince"—this sandwiched between "Let faith and truth give sureties of your life" and "use upright dealing both to man and wife" (sig. D3r). Either the state of the nation was nothing compared to the souls of the readers, or Churchyard decided to avoid a subject he could not safely and openly approach. The argument from lack of evidence is weak of course, but it is interesting to compare the ventriloquized "obey your Prince" with Twyne's lengthy placement of the national moment in his piece:

> 32. But shall wee now againe conjecture somewhat unto our owne comforte, and not altogeather unprobably? Since at all times these one and twentie yeares and upward, duringe the raigne of our most deere and dread Soveraigne, and most gratious Queene *Elizabeth*, the Gospell hath beene sincerely and truely preached unto us, and that now duringe this time of Lente last past, and since Easter, not only in her Majesties Courte, but also in her emperiall Cittie of *London*, as also in all other places of her dominions, most choice men for godlines and learninge have beene appointed to sow the seede of life, and to open the way unto the kingdome of Heaven: what if in token of consent, good liking, and conclusion of that which hath beene so manifoldly spoken, the Lorde would vouchsafe to give a nod with his head, wherat, as the holy Ghost speaketh by the mouth of the Prophet *David*: All the earth doth shake, and the hilles doo smoake, and the whole frame of the world is moved?
>
> (sig. B4r)

What if the earthquake is "a nod" from God, a gesture of "good liking" for the work of "the most gratious Queene *Elizabeth*" and her men, chosen for "godliness and learning"? This is, he says "not altogether unprobabl[e]" and is surely at least "somewhat unto our own comforte" to entertain. Here Twyne continues with his less than assertive style but his implication is grand, suggesting the quake is a sign not of trouble nationally but of national prosperity under good leadership. Twyne does turn directly to the more certain need for everyone to scan his and her own conscience, but his suggestion here leaves the door far more open to alternative readings in various directions than Churchyard's pamphlet deems safe.

The contrast between pamphlets reflects to us the readership at the time—a vast, growing, voracious, and well informed one that would continue to remember, imagine, discuss, and represent the quake in these ways and still many others. Twyne speaks to the reader who cannot dismiss natural arguments out of hand, who is more likely to do a bit of the intellectual "sift[ing]" Churchyard calls out as foolish, but who in the end also considers himself or herself grounded in Christian faith with its omniscient omnipotent God in charge. To this reader, Twyne offers materialist meteorology with a Christian and decidedly English Protestant frame, as William Fulke had done for meteorology by way of his *A goodly gallerye*. Twyne's account was not the last word on the 1580 earthquake, however. The Church of England was about to come forward with its official position.

*

Queen Elizabeth I and her Council were quite aware of the space opened by the earthquake for public pronouncements on cosmological, national, and local matters. They understood as well that events such as these always might be used to tell not only personal stories about "where I was when" the quake hit—as Juliet's nurse and Thomas Twyne do; they might also be used to narrate national history. Either this was a minor event, because the nation is generally on course, or this was an enormous event showing England is headed for destruction. Although the orders for prayer issued by Elizabeth I and her Council, by way of Bishop Edmund Grindal, do not appear in the Stationer's Register until June—and then they appear with a translated work on earthquakes by Abraham Fleming, also the subject of study in this chapter—Grindal had in April employed special prayers against the quake for London. Elizabeth I and her Council were meeting in April as well, to take these special prayers to the nation as part of increased church observance that would include fasting.

Such a practice of offering additional prayers was common by this time, with such orders for prayers issued during plague visitations and for other threats to national security, including the 1588 threat from the Spanish Armada. These were the Church of England's first special prayers for an earthquake but the haste in which the orders were created—already, as noted previously, in place in London by 22 April—is one reason we find a fair amount of overlap between the 1580 orders and the first ever nationwide special orders for prayer against the plague (*Church of England, A fourme to be used in common prayer twyse aweke, and also an order of publique fast, to be used every Wednesday in the weeke, duryng this tyme of mortalitie, and other afflictions, wherwith the realme at this present is visited. Set forth by the Quenes Majesties special co[m]maundement, expressed in her letters hereafter folowyng in the next page*). In 1563, when plague struck England for the first time in Elizabeth I's reign, it followed closely upon the queen's recovery from smallpox, and it left her councilors shaken. Her Council issued what was the first nationwide schedule of special prayers for plague, and they asked Alexander Nowell to contribute what would arguably be the most unique and important of pieces, the homily. The prayer schedule included psalms to sing, small meditations to say, requirements for fasting, and biblical passages to recite—all followed by the homily that serves as a sermon on God's justice and mercy.[11]

A comparison of the special prayers for plague, for the quake, and in the time of the threat from Spain would be worth lengthy undertaking, but here I call attention to the concern in the 1563 plague and

1580 earthquake prayers to justify fasting as an acceptable activity for Protestants and the degree to which both sets of prayers align Protestant England with Israel, and then use this relationship to remind God of the convent he has with these nations, thus meriting merciful fatherly treatment. Among the portions of the 1580 prayers that differ from those prior is the direction to householders for additional prayer at home and a separate prayer for the church. The special prayers' direction to families is "Also that they cause their family every night, before their going to bed, all together to say the prayer set out for that purpose, meekely kneeling upon their knees" (sig. A2v). The prayer in question is "A Prayer to be used of all householders, with their whole family, every Evening before they go to bed, that I would please God to turned his wrath from us, threatened in the last terrible earthquake" (sig. C4v-D1r). With it, the Church, Queen, and Council sought to unify people in practice, the better to increase at least the sense of national security.

England was then under siege by threats from Spain, incited by the Pope and a new Jesuit mission to England initiated in 1580, the very year in question. There was, on the other side of the religious divide, growing discontent among those who in their Protestantism leaned Calvinist and sought increased reform measures and to break free in many respects from all pre-Reformation rites and practices. On the benefit of changing twenty-first century thinking about human agency to that in which all "vibrant materials enter[ ] and leav[e] agentic assemblages," Bennett explains that it becomes possible to ask, "Can a hurricane bring down a president?" (107). Queen Elizabeth I and her Council knew that an earthquake could very well bring down a way of professing faith and perhaps even bring down a queen.

The 1580 earthquake prayers also, tellingly for the same reasons, include "A prayer for the estate of Christ's Church: to be used on Sundays"—a prayer not included among the plague prayers but which similarly aims for unity:

> Blesse all such (if it bee thy good will) whom though hast united and knitte unto us in any league of familiaritie or affinitie that we may rejoyce in the best bonde, and onley in this, that we are made partakers of one inheritance. Be mercifull unto thy people of Englande which confesse thy name. . . . Turn away thy wrath which thy terrible tokens do threaten toward us, and turn us unto thy selve, remove us not out of thy presence, but let thy fatherly warnings moove us to repentence.
> (sig. B1v–B2r)

The repetition of "us" and "we" as "thy people of England" is telling. So too, is "The Report of the Earthquake" that follows directly after the prayers for the church and before the homily. It is a report of the most significant of incidents from the quake but in concludes definitively that the quake it again showing this quake was "little above a minute of an houre, rather shaking Gods rod at us than smyting us according to our deserts" (sig. C1v). The message is intended to bring community minds and bodies back into prescribed union, into thinking and community sanctioned by the Church of England.

This intentional reinforcement of unity was a leading purpose for the creation of the full schedule of prayer—a claim certain in light the careful scripting of the purpose for these prayers by the Privy Council. Meeting on 22 April, the Council had decided it was time to weigh in with some force to commission from Edmund Grindal, archbishop of York and of Canterbury, just this nationwide schedule of prayer and fasting suited to this moment of religio-political need. Also Bishop of London, Grindal had already placed one into practice for London, and it was thought he might adjust these prayers for the larger audience and to state explicitly the Queen and Council's desires for it. On record in the Acts of the Privy Council is "A letter from the council to the Bishop of London touching the Fast and Prayers apointed sithe the Erthquake," the content of which follows:[12]

The Council to the Archbishop

> AFTER our very hearty commendations. Whereas an order of prayer and other exercises, upon Wednesdays and Fridays, to turn God s wrath from us, threatened by the late terrible earthquake, as an extraordinary token of his wrath against them, and fatherly admonition to turn them from their offences, and contempt of his holy word, as also of his infinite goodness and mercy to deal more favourably with us therein, than he hath dealt with other nations in the like case; in that we (thanks be unto his majesty!) have received no great hurt thereby, in comparison of that they have sundry times heretofore by the like occasion; whereby not only their houses and cities have been overthrown and destroyed, but also many thousands of peple have pitifully perished.
>
> And understanding that you have considered upon and appointed a good and convenient order of prayer, and other exercises to be used in all the parish churches of your diocese, upon Wednesdays and Fridays, for the turning of God his wrath from us, threatened by the said earthquake; with a goodly prayer for the like respect, to be used of householders with their families: we do not only commend and allow your good zeal therein, but also think the same very meet to be generally used in all other dioceses of this realm; requiring you to give order, that in every of the same the said whole some and godly order of prayer may, for the respect afore said, be executed, followed, and obeyed, during such time as you think meet. And so we bid your lordship most heartily well to fare.[13]

With this letter, the Privy Council not only instructs Grindal regarding the desired reach for the prayers—national—but also regarding the content of them, including how to interpret the earthquake: God uses earthquakes as a punishment, a sign of "his wrath" but England has merited also "fatherly admonition" intended to correct "their offences, and contempt of his holy word." God, they would have it believed, has "deal[t] more favourably with us therein, than he hath dealt with other nations in the like case." Compared also to those in "sundry times heretofore" who lost homes, cities, and lives, England "received no great hurt." The wording is interesting, aligning England and Israel, turning the quake into a relatively minor warning, and assigning blame only most generally to those in the nation with "their offences, and contempt of his holy word." England, the Queen and Council conclude, is in good stead and will continue so if there are a few small corrections made by individuals; these corrections can of course be aided if not entirely achieved by faithful attendance to these special, additional weekly prayers and fasting. London will be Nineveh but on a more modest scale than the rhetoric of Churchyard would demand.

The title and content of the homily commissioned for these special prayers reflect exactly the intention of the Council. For its composition, the Church selected not one of its own clergymen to perform the task, as it had formerly, but famed translator Arthur Golding. By this time Golding was already a well-respected translator both of Calvin's sermons and, as is better known today and figures in discussion in Chapter 1, of Ovid's *Metamorphoses*. With its display of a mother earth who quakes with passion and births monsters and men, Golding's Ovid is quite at odds with his Calvin as well as with his *A Godlie Admonition for the time present*, the homily in the Church of England's earthquake prayers treated below.

This homily has much in common with Bishop Aylmer's 1563 homily for the special orders for prayer in plague-time. It has also as much in common with the pamphlets of Churchyard and Twyne, but it balances the two treatments, as in its first lines:

> Many and woonderfull wayes (good Christian Reader) hathe God in all ages most mercifullye called all men to the knowledge of themselves, and to the amendemente of their Religion and conversation, before he have layd his heavy hande in wrathfull dyspleasure upon them. And this order of dealing he observeth, not onely towardes his owne deare children, but also even towardes the wicked and castawayes: to the intente, that the one sorte tourning from their former sinnes, and becomming the warer al their life after, shold glorifie him the more for his goodnesse in not suffring them to continue in their sinnes unreformed, to their destruction: and that the other sorte shoulde be made utterly unexcusable for their wylful persisting in the stubbornesse of their harde and frowarde heartes, against all his friendlie and fatherlie admonitions.
>
> (sig. c2r)

The earthquake is supernaturally sent, but it is not quite as Churchyard would have it be: it is not a sign of dire circumstances, of England careening off track. It is a modest call—memorable, certain, but not the moving of mountains, the sinking of towns. It is like those warnings through history given to "all men" generally, offering them fair and merciful reminders to have "knowledge of themselves," to scan their own consciences individually and determine what correction is necessary.

Still more to the point is Golding's placement of the quake in the context of England's history:

> Let us enter into our selves, and examine our time past. Since the sharpe tryall which GOD made of us in the raigne of Queene Marie, (at which time we vowed all obedience to GOD, if he woulde voutchsafe to deliver us againe from the bondage of the Romishe Antichryst, into the libertie of the Gospell of his sonne Jesus Chryste) he hearkening effectually to our requestes, hath gyven us a long resting and refreshing-time, blessed with innumerable benefites both of body and soule: For peace, health, and plentie of al things necessarie for the life of man, we have had a golden world above all the residue of oure neyghbours rounde aboute us. The worde of truth hath bin preached unto us earely and late without lette or disturbance. And bicause our prosperitie hath made us to play the wanton children against God, he hath chastized us in the meane season with many fatherlie corrections.
>
> (sig. F1r)

Here as in Twyne's account, England enjoys favored status in the eyes of God, with respect to her leadership. God heard England's prayers and delivered her from the reign of Mary Tudor, from "the bondage of the Romishe Antichrist," and to "libertie of the Gospell" with the result of "innumerable benefits" in a "golden world." The earthquake is not a sign that England as a whole is on the wrong track or that the head of the country or her ministers need correction; it is simply a matter of "fatherlie correction" because the people have indulged in their "prospertie" rather than persisting in precise faithfulness. All is well, Golding explains—with frequent mention of God's "fatherlie"ness; this is all a needful warning, and all will be well if the warning is heeded.

But Golding knew what all others did as well: not everyone was of the same mind in these matters, and the church was caught in the middle. What only was in common was the act of questioning. Particularly of concern were "some, which . . . will not stick to deface the apparent working of God, by ascribing this miracle to some ordinarie causes in Nature" (sig. E4v). What follows is Golding's effort then to expose as inapplicable these "ordinarie causes" for this quake. He does so, almost as Twyne had, perhaps drawing from his work. For example, he argues:

> whereas naturally Earthquakes are sayde to be engendred by winde gotten into the bowels of the earth, or by vapors bredde and enclosed

> within the hollowe caves of the earth, where, by their stryving and struggling of themselves to get oute, or being haled outwarde by the heate and operation of the Sun, they shake the earth for want of sufficient vent to issue out at: If this Earthquake had rysen of such causes, it coulde not have bin so universall, bicause there are many places in this Realme, which by reason of their substancial soundnesse and massie firmnesse, are not to be pierced by any windes from without.
>
> (sig. E3v–4r)

Golding delivers his materialist meteorology nicely, citing the theory in which wind comes from an external source or is bred within earth—either way struggling to get out. He even allows for some equivocal "or"s, but he nonetheless uses passive verbs and, like Churchyard, omits his sources, as if to diminish the authority of Aristotle, Seneca, and Pliny by not naming them. Golding also employs the terminology of "engend[ering]" in the "bowels" with "vapors bredde" such that they will be "stryving and struggling of themselves to get oute." Beneath the surface of his materialist theory deployed to Christian purpose is an account of the birth of Titans to a polytheistic mother earth. None of this is intentional, of course, and we are finding just the traces of this paradigm here, but it is noteworthy in relationship to what will come from the hand of Harvey later in this chapter and what we will see in those ensuing.

Golding remains firmly attached to proceeding through several additional examples of the misalignment between the 1580 quake and materialist quake theory, and then, toward the end of this set of anti-materialist proofs, he offers a gesture toward closure:

> we maye well conclude (though there were none other reason to move us) that this miracle proceeded not of the course of any naturall causes, but of God's only determinate purpose, who maketh even the very foundations and pillars of the earth to shake, the mountaynes to melte like Waxe, and the seas to dry up, and to become as drye fielde, when he listeth to showew the greatness of his glorious power, in uttering his heavy displeasure against sinne.
>
> (sig. E4r-F1v)

Indeed, "we maye well conclude" thus; but even after this statement, Golding dare not yet imagine a reader doing so. Instead, as Twyne had, Golding imagines a reader who is not yet satisfied:

> But putte the case that some naturall causes or secrete influences had their ordinarie operations in this Earthquake, whereof notwithstanding there is not any sufficient likelyhode: shall we so gaze upon the meane causes, that we shal forget or let slip the chiefe & principall

causes? Knowe we not (after so long hearing and professing of the Gospel) that a sparrow lighteth not on the ground without Gods providence?

(sig. F1v)

Earth itself, like the sparrow of Matthew 10:29, is subject in every particular and at all times to God's power, even when he works through nature. Letting natural explanations for earth's motion prevent or even distract one from turning to one's conscience is dangerous business. The only and ultimate "rocke, which neither wind, water, nor Earthquake, no nor Sathan himself with all his Feends can shake downe or empayere" is God (Sig F2v). And it is through the church that one will be, as here, reminded of "the time of oure visitation" so that one can "use it to our benefit" (sig. F3v). Still, Golding knew what a heavy burden of proof he carried in this effort. No other special prayers issued by the church under Elizabeth I include such experiential detail for events in question, and none supply conflicting causal arguments. These inclusions attest to the threat of this particular quake, to the influence of the press, and to threat of competing explanatory paradigms available for the interpretation of the quake. They also attest to the broad knowledge base, vividly particular vocabularies, and voracious imaginations of readers and listeners who were increasingly demanding far more than simple answers for complex events and situations.

*

Months later, the earthquake was still very much in mind, including for printing house translator, editor, indexer, and compiler Abraham Fleming (*c.* 1552–1607). He was already then an active lead editor of Holinshed's *Chronicles* (1577) when his work on the earthquake appeared in the Stationer's Register on 27 June.[14] It is the longest of all 1580 earthquake pamphlets and it is the oddest among the set. Here in full is its title: *A bright burning beacon, forewarning all wise virgins to trim their lampes against the comming of the Bridegroome. Conteining a generall doctrine of sundrie signes and wonders, specially earthquakes both particular and generall: a discourse of the end of this world: a commemoration of our late earthquake, the 6. of April, about 6. of the clocke in the evening 1580. And a praier for the appeasing of Gods wrath and indignation. Newly translated and collected by Abraham Fleming. The summe of the whole booke followeth in fit place orderly divided into chapters.* Unlike the other pamphlets, Fleming's is a translation of a pre-existing work on earthquakes. Fleming is also unique for following the meteorlogical theory not so much of Aristotle, Seneca, or Pliny but rather of Viennese Bishop Frederick Nausea (1496–1552). Nausea had labored unsuccessfully to engage Lutherans in a return to Catholicism, and was known for this. His thoughts on earthquakes, however, seem to be derived more from his own imagination and eclectic readings than from any Catholic interpretation of meteorology. Nausea writes passionately about earthquakes being sure signs of the literal illness and impending death of earth itself, which he calls "the mother of us all." These symptoms of illness point to the end days (sig. D1v).[15] Fleming, translating Nausea explains:

> Now there are diseases incident to the world, as we thinke: and what are they, but such effects as are denounced by signes and wonders to happen unto the world? namely, Earthquakes, overflowings of waters, fieres in the element, famines, pestilences, and any other of this sort: which are none otherwise foreshewed, either presently to happen, or hereafter to come to passe, than the sicknesses whereunto men are subject, and foretold them by certeine signes and tokens: as namely by their water, and other excrements.
>
> (sig. H4r)

Here is meteorophysiology in ample display, with all of earth's ailments aligned literally through matter and motion with diseases witnessed easily in the human body. Here too is a flattening out of individual differences among "diseases incident to the world" to render earthquakes, like "pestilences" and "and any other of this sort," exchangeable symptoms of illness.

What these many "diseases" show, as do "certeine signes and tokens" to which "men are subject," is the following, which places meteorology and physiology on even closer par with respect to the use of its symptoms toward diagnosis:

> For *Hippocrates*, to make the Physician cunning in foreknowing and also foretelling of such issues in the sicke, counselleth them to marke diligently the face and countenance. For if their eyes be hollow, their eares colde and shroonke together, their foreheade drie and withered . . . they shall be signes, that the patient will either be extreme sicke, and so escape verie narrowly, or else that there is no way with him but one, even death. To applie this to our purpose, when we see such wonders in the world, as are strange and fearefull, when we see them often in such maner as hath not beene in former times, it is an undoubted token that the world is not well, but infected with certein sore sicknesses, and like shortly to die, or else to fall into great danger: so much the rather, because signes and wonders, monstrous appearances, and strange sightes, have their generation herehence, either because particular nature faileth, or through the default of the matter it selfe which resisteth, or else by reason of the weaknesse of the agent or worker. Which is thought to be a token that the world is sicke after the maner of a man, who is therefore called a little world.
>
> (sig. I2v–I3v)

Fleming via Nausea reads the symptoms of the world's illness, seeing them "often in such maner as hath not beene in former times, it is an undoubted token that the world is not well." Individually and collectively, the events occurring are worse than those of the past, and this points to a "world . . . like shortly to die"—all part of God's coordinated plan for "*the comming of the Bridegroome*," as in the English title. Nausea refers as often, then, to Hippocrates as to Pliny and to Hippocrates more often than to Aristotle or Seneca. Using Christianized Galenic physiological theories to speak of the macrocosmos (in some ways as relentlessly as Helkiah Crooke would do for the microcosmos, treated in chapter 6), Fulke turns the universe into a body that God employs as a vehicle for human salvation.

Fleming concludes the translation and then offers an original, and the very first, catalog of English earthquakes—a task apt for him due to his work as editor of the Chronicles. The section he titles:

> A contemplation of wonderfull accidents, and principally of Earthquaks, as well particular as generall, which have happened in the realmes of England, Ireland, and Scotland, from the time of William Conquerour, to the reigne of our sovereigne Ladie and gratious Queene Elizabeth, &c. Also a commemoration of our late generall Earthquake the 6. of April, about 6. of the clocke in the Evening. 1580.
>
> (sig. M4r)

Among the earthquakes, however, he concludes the 1580 event was the worst of them. He asserts, "I may not so boldly as truly affirm, that the like was never heard of since the creation" (sig. N4r). In the context of the pamphlet's overt meteorophysiology, the 1580 earthquake is nothing less than earth experiencing asphyxiation, just as when:

> our own wind and breath stop[s] or sta[ys] in our breast, and not having recourse in and out by interchaunge of turns[;] we perceive & feele therby, that our very soule, or life being assalted, the limmes and members of our bodies are taken with a trembling, there is stirred up within us a kinde of strife or wrestling, & all the outwarde partes of our bodie, thorough feare fall a quivering: till this winde or breath hauling gathered force sufficient, find a way to avoid, and the pipes wherein it was kept burst open, it issue out with a vehement and great noise.
>
> (sig. B3v–B4r)

The 1580 earthquake is such a "burst[ing]" that—as he describes it in his chronology of English earthquakes—was:

> so universall (*for I beleeve the Lord did shake the foundations of the whole earth, & it was his mercie in that we were not all utterly*

> *undone*). I may conclude that it was supernatural, & being supernatural the more wonderful. For neither wind nor water could have the force, with a generall moving of the whole land, to terrifie the peoples hearts. Let us be resolved, that there remaineth nothing now, but the day of our visitation.
>
> (sig. G1v)

For Fleming, like Churchyard, the quake speaks of dire straits for England and of the need, now, for action by everyone. His pamphlet's vehemence outdoes that of Golding's church-sanctioned earthquake explanation. Indeed, the worst quake is under Elizabeth I, however "gratious" she is. Surely this is a pamphlet that kept quite alive the "grave Meteorologicall Conference"—so termed by the writer who penned the last of the 1580 earthquake responses.[16]

*

The credit for the closing remarks in the earthquake discussions of 1580 and for a turn in this examination to imaginative and predominantly literary treatments of earthquakes goes to that writer, Gabriel Harvey (1552/3–1631).[17] In a published letter exchange with Edmund Spenser—*Three proper, and wittie, familiar letters: lately passed betweene two universitie men: touching the earthquake in Aprill last, and our English refourmed versifying. With the preface of a wellwiller to them both*—Harvey takes a broad swipe at prior views on the earthquake, Christian- and materialist-leaning alike.[18] In the second letter of the three, in which he ostensibly responds to Spenser's request for news about the quake, Harvey recounts an evening at cards, beginning with an apology "for breaking one principall graund Rule of our olde inviolable Rules of Rhetorick, in shewing himselfe somewhat too pleasurably disposed in a sad matter" (9).[19] We know as readers that we are right away in very different representational territory. He proceeds then to explain that he was with his gentlemen friends and their "*shrewde wittie new marryed Gentlewomen*, which were more Inquisitive, than Capable of Natures works[ ]" (9) when the earthquake struck. Already we know that one of the purposes for this treatment of the earthquake that contrasts with all others is to produce humor. Harvey continues to set up the scene. The men had been playing cards prior to dinner. After the quake strikes, Harvey, among those card-playing gentlemen, sets a humorous tone, calling attention again, as will be the dominant rhetorical gesture, to the wives of the gentleman. They are in another room and Harvey queries, "Good Lorde, quoth I, is it not woonderful straunge that the delicate voyces of two so propper fine Gentlewoomen, shoulde make such a suddayne terrible Earthquake?" (10).

This initial setup and query works on several levels. It offers husband and wife banter, showcases convivial comfort among the company, and it draws the reader into the situation, helping him or her to imagine the scene. The joking about the voices of the women is also informed by materialist earthquake theory. The low roar that precedes an earthquake was thought caused by the rumbling of earth, but during the experience of hearing such an enormous, friction-produced roar, it can seem that the roar itself—caused perhaps by a biological creature, lion or human—is the cause of the movement. As Aristotle explains:

> Wind is also the cause of noises beneath the earth among them the noises that precede earthquakes, though they have also been known to occur without an earthquake following. For as the air when struck gives out all sorts of noises, so also it does when it is itself the striker. . . . The sound precedes the shock because the sound is of finer texture and so more penetrating than the wind itself . . . so that sometimes the earth seems to bellow as they say it does in fairy stories.
>
> (271, 219)

Aristotle draws what is intended to be a reasonable comparison between the "texture"-associated speed of sound versus that of motion. He implies that although the physical motion is the cause of the sound, and it is well to be informed about this and not mistake the situation such as to be surprised by the motion, the sound can function as a precursory warning of the motion. As Harvey's humor would have it, the further implication when related to marriage is that these men, and men in general, have had the experience of the gossiping or scolding of women as just such sound preceding active trouble. Compared to the pains to which prior writers on the quake went to specify whether sound came with the quake, as a matter of discerning its supernatural or natural cause, Harvey's treatment of this initial sound is unusual and engaging. According to Twyne, Churchyard, and the Church of England via Golding, no such sound was heard, though in natural quakes they typically are; thus they would say the quake was supernatural. Harvey does not attend to the sound as a matter of fact for deduction but rather it seems first for fun, either just for the story quality with its certain laugh at nagging wives, or for the satirical treatment of the earthquake pamphlet genre itself, with its seeming precision but uncertain and politically packaged claims.

Harvey continues, adding that he had not thought for a minute that there was anything especially strange going on when he heard the sound, because he was:

> Imagining in good fayth, nothing in the worlde lesse, than that it shoulde be any Earthquake in déede, and imputing that shaking to the suddayne sturring, and remoouing of some cumberous thing or other, in the upper Chamber ouer our Heades: which onely in effect

> most of vs noted, scarcely perceyuing the rest, béeing so closely and eagerly set at our game, and some of us taking on, as they did.
>
> (10)

Indeed, the statement about the wives' voices was purely for humor, yet the men nevertheless do not take the sound or the threat of a quake as a matter for much concern, "being so closely and eagerly set at our game." With this detail, Harvey speaks to the relative lack of fear that was felt by the men, and the degree to which it was likely that this quake went entirely unnoticed by many, as Twyne reports. The women, however, show concern, and Harvey's group decides to investigate, sending someone to ask about town: "we had by and by certayne word, that it was generall over all the Towne, and within lesse than a quarter of an howre after, that the very like be happened the next Towne too, being a farre greater and goodlyer Towne." This news increases the level of interest for the men and that of concern in the women, as he tells us:

> The Gentlewomens hartes nothing acquaynted with any such Accidentes, were marvellously daunted: and they, that immediately before were so eagerly, and greedily praying on us, began nowe forsooth, very demurely, and devoutely to pray unto God, and the one especially, that was even nowe in the House toppe, I beseeche you hartily quoth shee, let us leave off playing, and fall a praying.
>
> (10–11)

This call for prayer is a move that would in most prior pamphlets be recommended and lauded. But Harvey swiftly extends his poking fun of the women at every point, including at their request for prayer, at which he shows their husbands to scoff. Implying that they think the request for prayer to be a diversion from more important responses to the quake, they ask Harvey, the only university man present, to explain what it is that "Philosophers" might say regarding "some sensible Natural cause thereof, in the concavities of the Earth . . . some forceible and violent Eruption of wynd or the like?" (11). In their request, the gentlemen offer up additional points of materialist theory for the earthquake, which is perhaps a move by Harvey to show that non-university people already possess such "sensible" rather than superstitious understanding of quakes and other meteorological phenomena. Harvey initially responds to confirm and praise their learned inquiry, but true to form he does so by following that with diversionary humor:

> Yes no doubt, sir, may there, quoth I, as well, as an Intelligible Supernaturall: and peradventure the great aboundaunce and superfluitie of waters, that fell shortly after Michaelmas last, being not as yet dried, or drawen up with the heat of the Sun, which hath not yet recovered

> his full attractive strength and power, might minister some occasion thereof, as might easily be discoursed by Natural Philosophie, in what sorte the pores, and ventes, and crannies of the Earth being so stopped, and fylled up every where with moysture, that the windie Exhalations, and Vapors, pent up as it were in the bowels thereof, could not otherwise get out, and ascende to their Naturall Originall place. But the Termes of Arte, and verye Natures of things themselves so utterly unknowen, as they are to most heere, it were a peece of woorke to laye open the Reason to every ones Capacitie.
>
> (11)

Harvey displays his detailed Aristotelian reasoning regarding the moist earth as reason for "the poores, and ventes, and crannies of the Earth being so stopped" that earth's skin would form a barrier to the natural release of "windie Exhalations, and Vapors." All here supports a materialist reading of the earthquake—a reading challenging outright most that have come in the weeks following the 1580 quake. Yet Harvey ends with a move to elicit more conversation, all but challenging his listeners to query him further regarding the "verye Natures of things themselves so utterly unknowen," their causes or "Reason" beyond the "Capacitie" of many.

The gentlewomen will not allow this suggesting of their incapacity to understand to pass, however: "try our wittes a little, and let us heare a peece of your deep Universitie Cunning" they reply. It is a skillful device on the part of Harvey the author to have the gentlewomen demand still more of the character Harvey's learned knowledge on the subject while seeming to undercut "deep Universitie Cunning" at the same time. Harvey the character responds at length in what initially seems to be a serious answer to the gentlemen's query, reporting to set up the scene that he:

> then forsooth, very solemnly pausing a while, most gravely, and doctorally proceeded, as followeth. *The Earth* you knowe, is a mightie great huge body, and consisteth of many divers, and contrarie members, & vaines, and arteries, and concavities, wherein to avoide the absurditie of *Vacuum*, must necessarily be very great store of substantiall matter, and sundry Accidentall humours, & fumes, and spirites, either good, or bad, or mixte Good they cannot possibly all be, whereout is ingendred so much bad, as namely so many poysonfull, and venemous Hearbes, and Beastes, besides a thousand infective, and contagious thinges else. If they be bad, bad you must needes graunt is subiect to bad, and then can there not, I warrant you, want an Object, for bad to worke vpon. If mixt, which seemeth most probable, yet is it impossible, that there should be such an equall, and proportionable Temperature, in all, and singular respectes.
>
> (12)

The setup of this passage speaks volumes for Harvey's sense of himself as a character in the scene, of the role his friends ask him to play, associated with university learning, "solemnly . . . gravely . . . and doctorally." What he offers, however, is a blend of fable-quality polytheistic personification, with creatures inhabiting earth as "a mightie great huge body" replete with "vaines, and arteries" and Aristotelian rationality in procedure as if to suggest his content is indeed the product of university learning, when in fact it satirizes that learning.[20]

With the introduction of the "Evill (in the divels name)," the larger purpose of the work is obvious. This evil, he explains:

> will as it were interchaungeably have his naturall Predominaunt Course, and issue one way, or other. Which evill working vehemently in the partes, and malitiously encountering the good, forcibly tosseth, and cruelly disturbeth the whole: Which conflict indureth so long, and is fostred with aboundaunce of corrupt putrified Humors, and ylfavoured grosse infected matter, that it must needes (as well, or rather as ill, as in mens and womens bodyes) burst out in the ende into one perillous disease or other, and sometime, for want of Naturall voyding such feverous, and flatuous Spirites, as lurke within, into such a violent chill shivering shaking Ague, as even nowe you see the Earth have.
>
> (12)

Adding language from the Christian paradigm that includes a force like "Evill (in the divels name)," Harvey's odd amalgam tempers the power of such a figure by making its power only sometimes, "interchaungeably" with that of the good and mixed creatures, able to take the lead. When it does "have his natural Predominaunt Course," it still is a course that is not directed, as it "issue[s] one way, or other." In other words, any small advantage "evill" has is accidental and is, he then explains, tempered by "encountering the good" and jostling with it to "disturbeth the whole." An earthquake is the rare occasion when the conflict endures, grows, and, for a time, is impeded in what would regularly be "Naturall voiding" of the "corrupt putried Humors and ylfavoured grosse infected matter." In other words, the "matter" to be "voyd[ed]" might start benignly as "flatuous Spirits" commen to "mens and womens bodyes" but when held in, the result can be "violent," "as even nowe," he says, "you see the Earth have." It is as if the earth by this has plague and needs to push its venom out in the form of buboes; earth in this scenario undergoes "a violent chill shivering shaking Ague." This usage of "Ague" comes from the Latin word for "acute," associated both with something violent, sharp, and pronounced, and, for its early medical usage specifically, with "the initial stage of such a paroxysm, marked by an intense feeling of cold and shivering. Now chiefly *hist.*"[21] Thus Harvey—by way of a blended meteorophysiology, drawing from all three representational registers—explains the quake. This is in no way strictly the product of desiring to share one's university training or moral instruction. It is instead invitingly playful, even to the point of suggesting an earth suffering from intestinal distress.

Harvey explains further the ramifications of earth's illness, specifically explaining that earth's

> Ague . . . we schollers call grossely, and homely, *Terrae motus*, a moving, or sturring of the Earth, you Gentlewomen, that be learned, somewhat more finely, and daintily, *Terra metus*, a feare, and agony of the Earth: we being onely moved, and not terrified, you being onely in a manner terrified, & scarcely moved therewith.
>
> (12–13)

Harvey's observation recalls the initial reactions of the gentlemen versus gentlewomen—the former wanting reason and the latter wanting prayer. What we see, then, is a developing comment on the degree to which that people respond differently to wonders. One size will not fit all. For some, the quake is a *Terrae motus* and for others a *Terra metus*, for some movement and for others fear. What then of the need to argue for one approach over another? In answer, Harvey seems to dilate his thinking process, taking his audience into the uttermost forms of absurdity, testing the limits of knowledge and of his audience's patience:

> Nowe here, (and it please you) lyeth the poynt, and quidditie of the controversie, whether our *Motus*, or your *Metus*, be the better, & more consonant to the Principles and Maximes of Philosophy? the one being manly, and devoyde of dreade, the other woomannish, and most wofully quivering, and shivering for very feare. In sooth, I use not to dissemble with Gentlewoomen: I am flatly of Opinion, the Earth whereof man was immediately made, and not woman, is in all proportions and similitudes liker us than you, and when it fortuneth to be distempered, and diseased, either in part, or in whole, I am persuaded, and I believe Reason, and Philosophy will bear me out in it, it only moveth with the very impulsive force of the malady, and not trembleth, or quaketh for dastardly feare. Nowe, I beseeche you, what thinke ye, Gentlewomen, by this Reason?
>
> (13)

Again, Harvey changes the conversation topic as he proceeds through his explanation. Initially the question was regarding why the earth shakes and what are the details regarding natural factors that would cause it to do so. Then it becomes a consideration of male versus female reactions to the phenomena of shaking, and prior to that of hearing before feeling the quake. Next, we arrive at the most absurd jolt: the suggestion that—as borne out by "Reason, and Philosophy"—earth is male rather than female. In this *reductio ad absurdum*, Harvey aims for laughs not only at the content of his ridiculous conclusion that earth is more like man than a woman but also for laughs at those who in prior pamphlets and with urgency and earnestness put forward equally untenable claims.

The gentlewomen, who are intelligent and curious contrary to Harvey's implications, call him out on just this rhetorical adventuring:

> I can neither picke out Rime, nor Reason, out of any thing I have heard yet. And yet me thinks all should be Gospell, that commeth from you Doctors of Cambridge. But I see well, all is not Gold, that glistereth. In deede, quoth Mistresse *Inquisitiua*, here is much ado, I trowe, and little helpe. But it pleaseth Master *H.* (to delight himselfe, and these Gentlemen) to tell vs a trim goodly Tale of Robinhood, I knowe not what.
>
> (13)

By this time in Harvey's explanation for the earthquake's cause, a response of "I knowe not what" is most apt. Whatever Harvey is doing, it is the equivalent of telling some tall tales, perhaps based on some truth, perhaps not. Later the gentlewomen will interrupt Harvey again: "No more Ands, or Ifs, for Gods sake" (15). They delightfully track and react to his foolery but never divert him from his own entertaining diversions.

In response to the women's "Robinhood" accusation, Harvey replies that, after all, he *had* told them they might not understand his University learning—a response that he then again undermines with his most absurd move yet:

> if ye wyll give me leave upon that small skill I have in Extrinsecall, and Intrinsecall Physiognomie, & so foorth, I will wager all the money in my poore purse to a pottle of Hyppocrase, you shall both this night, within somwhat lesse than two howers and a halfe, after ye be layed, *Dreame* of terrible straunge Agues, and Agonyes as well in your owne prettie bodyes, as in the mightie great body of the Earth.
>
> (14)

Harvey teases again with the women, as if to regain their favors. At the same time, he reinforces the notion of a sympathetic, even empathetic, relationship between human bodies and that of earth when he says that his description of earth's illness will lodge so potently in the listeners' imaginations that it will breed there and conjure dreams in which the listener becomes the earth itself shaking in illness. Humor again is the leading purpose of this rhetorical move, but Harvey also means to suggest to the women that his "Robinhood" is a story with undeniable psycho-physiological resonance.

Sensing he has a captive audience, Harvey offers more and still more elaborate explanations for the shaking of the earth, each one the result of the request for his university-educated opinion. Some "learned" people, for example he explains, "defend the position . . . that the Earth having taken too much to drinke and as it were over lavish cups . . . now staggereth and reeleth and tottethereth" (14). Or it might be that, as in the "Ague" description, creatures war internally within earth, just as humans war on the surface of the earth, and they do so even until "one partie bendeth all the force of his Ordinance and other Martiall furniture against the other," which shakes the earth (15). Animal fables follow a drunk earth, both explanations of the earthquake supplying fantastic qualities surpassing those in any Robinhood tale. All options, intentionally preposterously, also nevertheless and as noted, he draws from key components of available cosmological paradigms. In the first instance, earth is polytheistically alive, shaking not to teach humans a lesson but rather as the result of having simply had too much to drink. In the second instance, war, "Gunnes" (15), and earthquakes combine due to the choleric heat of animals stirring in earth—the very terminology used by Shakespeare to describe the men at Harfleur and, as discussed in chapter 4, to describe some of the combatants in the War of the Roses. In each case, the instances reinforce the notions that earth and her creatures are no different than humans; that whatever the reason for the shaking, there is little of substantial meaning to be drawn from it; and that humans in particular are not part of the equation at all, nor is England or a Christian God.

Harvey again in this move stands apart from all other writers on the quake who never question the degree to which humans, as God's creatures, are involved and impacted by the quake. Harvey keeps much closer company in this with Margaret Cavendish, writing decades later of *The Blazing World* and its subterranean inhabitants, and here, with these words in "A World in an Earring":

> An *Eare-ring round* may well a *Zodiacke* bee,
> Wherein a *Sun* goeth round, and we not see.
> . . . .
> There *nipping Frosts* may be, and *Winter* cold,
> Yet never on the *Ladies Eare* take hold.
> And *Lightings*, *Thunder*, and great *Winds* may blow
> Within this *Eare-ring*, yet the *Eare* not know.
> . . . .
> There *Earth-quakes* be, which *Mountaines* vast downe fling,
> And yet nere stir the *Ladies Eare*, nor *Ring*.[22]

No sin or good behavior on the part of humans makes one jot of difference to the inhabitants of the Eare-ring, or vice versa. Although Harvey's eruptive animals in earth shake the surface enough to jostle the humans, causing a quake, Harvey—unlike the others writing on the 1580 quake—has already presented readers with a scene showing the quake to be so little as to have seemed only like the rumble of furniture being moved around on a second floor. Much ado about nothing; this is the earthquake of 1580 for Harvey.

Harvey's gentlemen friends soon again intercede, playing the earnest reader who really had hoped for some overtly serious and more rationally satisfying answers regarding the quake. They ask Harvey to take up their original question and let them know whether the quake was supernaturally or naturally caused and why. A rare section heading appears at this point in the letter, signaling "Master H^{s}. short, but sharpe, and learned Judgement of Earthquakes." In answer to the question, finally, Harvey covers one at a time, in roughly one full quarto page each, the Aristotelian four causes for the quake. As for the first:

> The Materiall Cause of Earthquakes, (as was superficially touched in the beginning of our speache, and is sufficiently proved by *Aristotle* in the seconde Booke of his *Meteorology*, is no doubt great aboundance of wynde, or stoare of gross and dry vapours and spirites, fast shut up, & as a man would saye, emprysoned in the Caves, and Dangers of the Earth which winde, or vapors, seeking to be set at libertie, and to get them home to their Naturall lodgings, in a great fume, violently rush out, and as it were, breake prison, which forcible Eruption, and strong breath, causeth an Earthquake. As is excellently, and very lively expressed of *Ovid*.
>
> (16–17)

Harvey shows the working together of materialist and polytheistic theories—the one "sufficiently prooved" and the other "very lively expressed" and therefore memorable, such that he goes on to quote the lines "of *Ovid*" from memory. Harvey provides the Latin version of Ovid, and here following is Golding's translation:

> Not farre from Pitthey Troyzen is a certeine high ground found
> All voyd of trees, which heretoofore was playne and levell ground,
> But now a mountayne for the wyndes a (woondrous thing too say)
> Inclosed in the hollow caves of ground, and seeking way
> Too passe therefro, in struggling long too get the open skye
> In vayne, (bycause in all the cave there was no vent wherby
> Too issue out,) did stretch the ground and make it swell on hye,
> As dooth a bladder that is blowen by mouth, or as the skinne
> Of horned Goate in bottlewyse when wynd is gotten in.
> The swelling of the foresayd place remaynes at this day still,
> And by continuance waxing hard is growen a pretye hill.
>
> (fol. 190r–v)

Here again then is materialist meteorological theory and another sub-stratifying layer of polytheistic meteorological theory that together have in Ovid's expression of them enchanted Harvey. This explanation for the formation of a mountain makes such an impression on Harvey that he mentions that memory alone is his source for the citation. This account of the material cause of the quake is coupled, then, with a literary addition that challenges strictly materialist claims for the quake. It will also surely challenge the forthcoming claims Harvey makes for any final cause of the quake.

On the final cause for the quake, however, he offers his most involved and earnest answer, which to quote in full is impossible. In sum, he asserts in order, that to try to determine a final cause other than that "the wynde shoulde recover his Natural place . . . no not our excellentest profoundest Philosophers themselves" would do (17). Sometimes, however, he says next, that of course God can bring about miracles. Still, then, because God works through natural causes, he need not often count on miracles and to for humans use natural events to prognosticate is wrong-headed and prideful (18). Going still one step further than all others writing directly on the 1580 earthquake, he says pointedly:

> But yet, notwithstanding, dare not I aforehand presume thus farre, or arrogate so much unto my selfe, as to determine precisely and peremptorily of this, or every the like singular Earthquake, to be necessarily, and undoubtedly a supernaturall, and immediate fatal Action of God, for this, or that singular intent, when as I am sure, there may be a sufficient Naturall, eyther necessarie or contingent Cause in the very Earth it selfe. . . . To make shorte, I cannot see, and would gladly learne, howe a man on Earth, should be of so great authoritie, and so familiar acquaintance with God in Heaven, . . . to reveale hys incomprehensible mysteries, and definitively to give sentence of his Majesties secret and inscrutable purposes.
>
> (19)

Harvey maintains the most moderate of Christian views here. The quake occurred due to natural matter and motion behaving as it should, as God generally designed. He joins *Anti-Prognosticon* and *A goodly gallerye* writer William Fulke in crying out against those who are for him the fools: not Churchyard's fine wits who sift their senses, but men like Churchyard who are "of so great authoritie, and so familiar acquaintance with God in Heaven" as to pinpoint precisely each message God sends in each separate earth shaking. This is to him as preposterous as any Robinhood tales.

Such prophesying might also be harmful, according more directly to Fulke, who says in his *Antiprognosticon*:

> what confidence hathe be[en] in [G]od or his worde, that dare not take in hande any honest and vertuous affaires (in which God

> hath promised to ayde and set forwarde all them that love hym) except e must fyrste aske counsayle of a blynde southsayer and Astrologer?[23]

For Fulke this is a matter of faith indeed, to separate himself from those who would do otherwise than trust in the natural working order of the universe created by a God who obeys his own word, who inscribed that word in the book of nature.[24] What a greater wonder, then, Harvey notes as a last word on the subject, is:

> the staying and quieting of the Earthe, beeing once a moving? May it not seeme a more myraculous woorke, and greater woonderment, that it shoulde so suddainely staye againe, being moved, than that it shoulde so suddainely moove, beyng quiet and still? Moove or turne, or shake me a thing in lyke order, be it never so small, and lesse than a pynnes Head, in comparison of the great mightie circuite of the earth, and see if you shall not have much more a doo to staye it presently, beeing once sturred, than to sture it at the very first.
>
> (22)

This conclusion delights both the gentlemen and gentlewomen, and the group advances to dinner.

This conversation was, for the dining companions as for the reader, one to remember. Harvey later thus dubs it "our yesternyghtes grave Meteorologicall Conference" (23). Whereas the turning of *meteora* into interchangeable billboards for propaganda flattens each event and each representation of it, making the accounts nearly as interchangeable as the events they narrate, Harvey's earthquake letter preserves multiple meanings, generating stories that excite the imagination and make a lasting impression on readers. Harvey illustrates the features of memorable impression by way of one last word on the "Meteorologicall Conference": the group of friends—gentlemen, gentlewomen, and Harvey—retires to dinner, where, Harvey says:

> our Supper put the Earthquake in manner out of our myndes, or at the least wise, out of our Tongues: saving that the Gentlewoomen, nowe and then pleasauntly tyhyhing [tee heeing] betweene them selves, especially Mystresse *Inquisitiva*, (whose minde did still runne of the drinking, and Neesing of the Earth,) repeated here, and there, a broken peece of that, which had been already sayde before Supper.
>
> (23)

The powerful representation of a living earth—a drunk and sneezing one, no less—triumphs among explanations. It will be the impression that lasts, or at least the one Harvey would like to have at the last, by his mentioning of it in this humorous way again here. His fine materialist meteorology and his assurance that God is nevertheless the ultimate cause of the quake are no match for the image of a drunk and sneezing mother earth. The earth, female or male as Harvey might have it, is undeniably the generative center of this story—unless that is, one would want to count Harvey the author-character himself. A small, delightful and etymological piece of evidence for this claim is the Oxford English Dictionary's citation of Harvey's "tyhying" [sic] as the first use of "tee hee" as a verb in the English language (s.v. tee hee, noun; s.v. "tee hee," derivative verb [intr.]). The ladies' tee heeing, which the OED describes as "to utter *tee-hee* in laughing; to laugh affectedly or derisively; to titter, giggle," is for the dinner party lasting impression of the 1580 earthquake. A giggle of joy, of near embarrassed delight perhaps, is a remarkable response to an earthquake. A giggle can be a recalcitrant challenge to the status quo, even if it is not Titanic in proportion or other appreciable measure. Harvey's giggle is nevertheless important, because it can be read diversely, not only as a way of showing the women to be naïve. We might indeed see Harvey as belittling his fictional gentlewomen, but I read this pamphlet as showing through their response to Harvey's lessons the power of forthright and open-minded inquiry and the power of a healthy dose of laughter in response to authoritative claims that are absurd.[25] Harvey's gentlewomen thus help him satirize the efforts of the male writers who so earnestly and in some cases condescendingly took it upon themselves to school England the proper response to the 1580 earthquake. The earthquake tee hee is purposely both improper and appropriate; it is also indelibly memorable.

The last entry in the register for 1580 with "earthquake" in its title appears 6 October and is an entry reporting "earthquakes" among many "wonders" in Rome.[26] Four years later, in June of 1584, there is notice of an earthquake in Geneva. Earthquakes appear in the register only every few years after this and always in other countries.[27] The earthquake of 1580 was the exception in British history, among the largest of quakes and more notable than others also because print placed it into still wider circulation. There was no other single wondrous event—monstrous birth, tempest, comet, or plague outbreak—that had such pronounced, immediate, and sustained print response. There was no other single event that garnered such wide interpretation, its range astounding. The works above also speak polyvocally to various groups, even when not entirely intended to do so. Each also figures as part of the complex phenomenon of the 1580 earthquake—the textual aftershocks of the assemblage of human and non-human, animate and inanimate actants. The 1580 earthquake experience itself is in this way a recalcitrant Titan, with its own eruptive and changing forces that will unmistakably resemble those placed into action by Edmund Spenser, the subject of the next chapter.

Notes

1. Colin S. Harris, Malcolm B. Hart, Paul M. Varley, and Colin D. Warren, eds., *Engineering Geology of the Channel Tunnel* (London: Thomas Tellus, 1996), 196; R.M.W. Musson, "A History of British Seismology," *Bulletin of Earthquake Engineering* 11.3 (2013). See also Alok Jha, "London Is Overdue for a Major Earthquake, Warns Seismologist," *The Guardian* (16 September 2010; http://gu.com/p/2jmca/sbl, accessed 31 January 2016).
2. For example, Churchyard reports,

   > A number being at the Theatre and the Curtaine at Hollywell, beholding the playes, were so shaken, especially those that stoode in the hyghest roomes and standings, that they were not a little dismayed, considering, that they coulde no waye shifte for themselves, unlesse they woulde, by leaping, hazarde their lives or limmes, as some did in déede, leaping from the lowest standings.
   >
   > (*A warning for the wise* [1580], sig. B2r)

   See also Fleming, who names both theatres as especially sinful places inciting God's anger at England:

   > Doth not God sée your filthines, or thinke you that your trade of life depending wholy upon those your Heathenish exercises, are not offensiuve to his Majestie? Will he winke at such wickednes, & kéepe silence at such filthines as is continually concluded upon and committed in your Theatre, Curtaine, and accursed courtes of spectacles?
   >
   > (*A bright burning beacon* [1580], 15–16)

3. Todd A. Borlik, *Ecocriticism and Early Modern English Literature: Green Pastures* (New York: Routledge, 2010), 64.
4. For more on Nowell, see Stanford Lehmberg, "Nowell, Alexander (*c.*1516/17–1602)," *Oxford Dictionary of National Biography* (Oxford: Oxford University Press, 2004; online edn., January 2008, www.oxforddnb.com/view/article/20378, accessed 10 March 2013). It is interesting that in spite of these claims, there is no record that Churchyard was of "Gentleman" status, and his various claims about himself in multiple writings make it hard to assess the truth Raphael Lyne, "Churchyard, Thomas (1523?–1604)," *Oxford Dictionary of National Biography* (Oxford: Oxford University Press, 2004; online edn., May 2006, www.oxforddnb.com/view/article/5407, accessed 7 May 2016).
5. For an examination of Churchyard's account of the 1578 Norwich Pageant, considered typical of Churchyard's writings in that it includes bold claims for authority and an unusual amount of autobiographical detail for the time, see David Bergeron, "The 'I' of the Beholder: Thomas Churchyard and the 1578 Norwich Pageant," in Jayne Elisabeth Archer, Elizabeth Goldring, and Sarah Knight, editors, *The Progresses, Pageants, and Entertainments of Queen Elizabeth I* (Oxford: Oxford University Press, 2007), 142–159.
6. Kirilka Stavreva, "Fighting Words: Witch-Speak in Late Elizabethan Docu-Fiction," *Journal of Medieval & Early Modern Studies* 30.2 (Spring 2000): 309–338.
7. Norman Moore, "Twyne, Thomas (1543–1613)," rev. Rachel E. Davies, *Oxford Dictionary of National Biography* (Oxford: Oxford University Press, 2004; www.oxforddnb.com/view/article/27927, accessed 21 February 2016). Among Twyne's works in the years immediately preceding the 1580 pamphlet, see *Orbis terrae descriptio. English: The surveye of the world, or situation of the earth, so muche as is inhabited. Comprysing briefely the generall partes*

thereof, with the names both new and olde, of the principal countries, kingdoms, peoples, cities, towns, portes, promontories, hils, woods, mountains, valleyes, rivers and fountains therin conteyned. Also of seas, with their clyffes, reaches, turnings, elbows, quicksands, rocks, flattes, sheles and shoares. A work very necessary and delectable for students of geographie, saylers, and others. First written in Greeke by Dionise Alexandrine, and now englished by Thomas Twine, Gentl. (Imprinted at London: By Henrie Bynneman, Anno. 1572); *The wonderfull woorkmanship of the world: wherin is conteined an excellent discourse of Christian naturall philosophie, concernyng the fourme, knowledge, and use of all thinges created: specially gathered out of the fountaines of holy Scripture, by Lambertus Danaus: and now Englished, by T.T.* (Imprinted at London: [By John Kingston] for Andrew Maunsell, in Paules Church-yard at the signe of the Parret, 1578); and *A view of certain wonderful effects, of late dayes come to passe: and now newly conferred with the presignyfications of the comete, or blasing star, which appered in the Southwest upo[n] the x. day of Novem. the yere last past. 1577. Written by T.T. this .28. of November. 1578* ([Imprinted at London: By J. Charlewood for] Richarde Jones, and are to be sould over against Saint Sepulchres Church without Newgate, 1. December 1578).

8. Twyne's translation of Petrarch is also of interest. In this work, the character Joy would like to find something in life that offers stability and a sense of comfort in a changing world—such as that "the earth wyl stand steddy under foote." Joy is countered by the character Reason, who point outs the problem with all such claims for assurance. For example, Joy speaks of "steddy ground" as something upon which humans can depend, and Reason replies,

> But many tymes it hath not stoode, and for confirmation hereof, I let passe auncient examples, as *Achaia*, and the residue of *Greece*, with *Syria* and other countreis, where in tymes past both whole Cities have ben utterly swalowed up, and hilles sunke downe, & Ilandes drowned: . . . Thou hast seene Cities, strong Castls, and Townes, at one tyme standyng most firmely, which within few dayes after, a miserable and feareful sight, lay al flat upon the earth. . . . And therfore ceasse thou to be carelesse where is no securitie.
>
> (sig. P2r-P3v)

Their responsive dialogue illustrates that between Stoic and secular versus Christian attendance to all aspects of bad fortune.

9. R.M.W. Musson, "A History of British Seismology," *Bulletin of Earthquake Engineering* 11.3 (2013): 715–861, 725.
10. On this trope associated with the epic, see Rebecca Totaro, *The Plague Epic in Early Modern England: Heroic Measures, 1603–1721* (New York: Routledge, 2012).
11. On these prayers, see Totaro, *The Plague in Print: Essential Elizabethan Sources, 1558–1603, Medieval & Renaissance Literary Studies* (Pittsburgh: Duquesne University Press, 2010), 17–48.
12. John R. Dasent, *Acts of the Privy Council of England* (London: Printed for H.M.S.O. by Eyre and Spottiswoode, 1890), 450.
13. Edmund Grindal and William Nicholson, *The Remains of Edmund Grindal, D.d: Successively Bishop of London and Archbishop of York and Canterbury* (Cambridge, UK: Printed at the University Press, 1843), 416–417.
14. Cyndia Susan Clegg, "Fleming, Abraham (*c.*1552–1607)," *Oxford Dictionary of National Biography* (Oxford: Oxford University Press, 2004; www.oxforddnb.com/index/9/101009693/, accessed 6 July 2016).

15. Fleming is also the only writer of pamphlets to offer a list of the other writers "whose reports of our late Easter Earthquake, &c. are printed and published," suggesting that those who contributed pamphlets understood, at least after a point, that they were entering a discussion if not a debate. Of the nine writers named—and of which Fleming himself is the ninth—only four others have works known to us: Tarleton (via Churchyard), Churchyard, Twyne, and Golding. The other four penned works no longer exist, have not been identified, and/or were not entered in the Stationer's Register: works by Francis Shackleton, John Philippes, Robert Gittins, and John Grafton. My findings indicate that there are no records related to these men beyond their names associated with the earthquake here. Of the four men, none can be identified; nor can we identify any of their texts with certainty. One possibility conjectured by the editors of the Stationer's Register is that the entry for 8 April to be printed by H. Kyrkham (Lycenced unto him under th[e h]and of the wardens *A fatherly admonycon and lovinge warnynge to England but especially to London by the reason of a moost fearfull earth quake which he sent as a fayre token of his spedie commynge to Judgement: the vi of Aprill* 1580) was written by one of these authors. According to John Payne Collier, "This tract has not reached our time, but it may have been, and probably was, one of the nine which were, as Abraham Fleming tells us in that he produced on the occasion, written by different authors" (John P. Collier, *Extracts From the Registers of the Stationers' Company: Of Works Entered for Publication Between the Years 1557 and 1570, With Notes and Ill* [London: Shakespeare Society, 1848], 114–115). It is interesting to imagine the authorship and content for the nearly dozen registered earthquake texts that we cannot identify beyond their Register entry. Fleming gets us closer than any other source to being able to do so.
16. Harvey in Edmund Spenser, *Three proper, and wittie, familiar letters: lately passed betweene two universitie men: touching the earthquake in Aprill last, and our English refourmed versifying With the preface of a wellwiller to them both* (1580), 23.
17. For more on Harvey, see Jason Scott-Warren, "Harvey, Gabriel (1552/3–1631)," *Oxford Dictionary of National Biography* (Oxford: Oxford University Press, 2004; online edn., January 2016; www.oxforddnb.com/view/article/12517, accessed 7 May 2016).
18. Passannante examines Harvey's larger objective with the letters as being the taunting of conservative translator Arthur Golding by way of Ramist and Lucretian theory, with which he may have been familiar ("The Art of Reading Earthquakes," 792–832). See also Gerald Snare, "Satire, Logic, and Rhetoric in Harvey's Earthquake Letter to Spenser," *Tulane Studies in English* 18 (1970): 17–33.
19. It is not entirely clear that these letters were actually exchanged between the men in advance of their publication or the degree to which Spenser has a hand in them at all. They may be Harvey's fabrication, with Spenser's name used because they are friends and it might delight him but also because Spenser was the more appreciated of the two writers.
20. On the changing nature of "mixture," increasingly appreciated particularly in politics over time, see Wolfram Schmidgen, *Exquisite Mixture*.
21. OED, s.v. acute, noun, see the etymology; s.v. ague, noun, 1 and etymology.
22. Margaret Cavendish, "A World in an Earring," in Kate Lilley, editor, *The Blazing World and Other Writings* (New York and London: Penguin Classics, 1994).
23. See Fulke's *Antiprognosticon*, Sig. D3v.

24. On the book of nature as the word of God in this period, see James J. Bono, *The Word of God and the Languages of Man: Interpreting Nature in Early Modern Science and Medicine* (Madison: University of Wisconsin Press, 1995).
25. Winfried Schleiner is absolutely right to call this letter "the most gendered early modern response to earthquakes" ("Early Modern Recovery: Harvey's Gendered Response to an Earthquake in Essex, England, on 7 April 1580," *Cahiers Elisabethains* 70 [Autumn 2006]: 15) and to note that "Harvey is unusual in repeatedly tying the explanation of earthquake to the topic of gender difference" (17), but Harvey does little overly with this other than for humor. On the one hand he exaggerates, and on the other diminishes gender difference, showing the women more inquisitive than the men even if he ridicules them (and for no reason, because his words are the ridiculous ones that they rightly call out. I am under these circumstances still looking for a way to distill my own findings across the readings of texts for this volume with respect to gender. In other words, here and in chapter 4 on cursing, it is not clear that gender matters, in spite of its appearing to do so on the surface. See my notes on Suffolk, Richard, and Margaret, for example. See also on the history of these issues Carolyn Merchant, *The Death of Nature: Women, Ecology and the Scientific Revolution* (New York: Harper and Row, 1983); Mary D. Garrard, *Brunelleschi's Egg: Nature, Art, and Gender in Renaissance Italy* (Berkeley and Los Angeles: The University of California Press, 2010); and D.J. Haraway, *Simians, Cyborgs, and Women: The Reinvention of Nature* (New York: Routledge, 1991).
26. *A Transcript of the Registers of the Company of Stationers of London*; 1554–1640 A.D. Volume II (London, 1875), 173.
27. October 6 1580: Edward white Lycenced unto him under the. Bishop of LONDONS hand *the descriptyon of great wonders scene the xiii of January* 1580 *and fearfull wyndes and earthquakes at Roome*, June 23 1584, Thomas Woodcock, Licenced unto him under master Watkins handes, *A true discourse happened by an earthquake primo mariii* 1684, *in the places adjoyninge to the lake of Geneva . . . 26 Dec* [1590]. John Wolfe, Entred unto him for his Copies under the handes of Master Hartewell and Master warden Cawood, A booke intituled *a true Discripcion of the fearefull yearth quake which happened at Vienna in Austria on the xv daie of September* 1590 . . . John Wolfe / *Item a ballad discribinge the same Cittie of Vienna together with the yearth quake.*

3 Much Enmoved, but Steadfast Still Persevered

The 1580 earthquake literally rumbled early modern English bodies, and just enough of them to become the talk of the nation. English writers took advantage of the wide opening they had to insert their ideas on the quake into the related conversations occurring in English homes—all thanks to the printing press. Those writers knew the power of such a textual move of self-presentation. The earthquake could be a vehicle for their own social advancement. They also recognized the potential for this rare phenomenon to register with epic proportion in English imaginations, and that it had provided them an opportunity to claim some of the territory of literary history that their Continental neighbors had long controlled. In literature expressly, the subject of this and the following chapters, writers flexibly and creatively combined aspects of the three primary cosmological paradigms treated in chapter 1. They did so with liberty not as comfortably exercised by the writers treated in chapter 2—writers of the 1580 earthquake pamphlets and other texts explaining meteorological phenomena. Those writers carried the burden of faith, which demanded that earthquakes show themselves as flat signs of God's judgment and mercy. The writers venturing to represent the earthquake with entertainment rather than exhortation as their purpose instead attended devotedly to the demands of their readers and audiences who thrilled over stories of eruptive meteorophysiological alteration. In this endeavor, these writers gave themselves room to play, to invent, and to appropriate even the direst of natural events that could lead to disaster, such that these events would serve decidedly literary turns. In some respects more than the religious writers, these writers tamed the earthquake and the volcano, the tempest and the flood, coaxing them into shapes that nature herself, they might say, had not dared offer. The religious writers in contrast were limited, forced to present two-dimensional, biblically referential and deferential examinations that could never compete seriously in an effort to excite discussion. Only Gabriel Harvey's 1580 earthquake response, for example, could elicit a reader's tee heeing.

Among those attentive to the power of such stories of epic wonder over imaginations and over entire nations was Harvey's close friend, Sir Edmund

Spenser. Seeking to secure the favor of Queen Elizabeth I by giving her a national epic all for herself, Spenser fashioned a work in which it might seem that the very forces of the sublunary sphere appeared and reconfigured themselves solely for the purpose of serving England's sovereign. To accomplish his feat of bending matter and motion to serve her, Spenser set out on a multi-book project that he left unfinished but that along the way won him her recognition and national fame. Key in this success are his characters, the heroes and monsters of his tale, all of whom are animated by meteorophysiologically wondrous qualities. The focus of this chapter will be on the meteorophysiological qualities of the earthquake and its early modern kin, the volcano and tidal wave—qualities which contrast with those of the metal-stone adamant. This contrast is a matter of scale and time, as the monsters of earthquake, tsunami, and volcano are large but short-lived and the heroes of adamant are comparatively small but resilient. This range, from earthquake to adamant, large to small but also short-lived to steadfast, speaks to the meteorophysiological condition of change—on one end massive, sudden, disruptive change that cannot endure, and on the other almost imperceptible because slow and small change that can. Spenser's Orgoglio and his dragon of book 1 are such highly changeable, massive monsters, and they stand in contrast to Spenser's most powerful-because-steadfast knights, Arthur and Britomart.

*

With Orgoglio, Spenser initiates his treatment of the Titans who will rumble through the entire epic, ever challenging the stability of Faerie Land and of civilization writ large. Little wonder it is that Orgoglio is the first of Redcrosse's enemies able to best him in combat; both Titan and earthquake by birth, Orgoglio is a formidable foe, as Spenser explains:

> The greatest Earth his uncouth mother was,
> And blustring *AEolus* his boasted syre,
> Who with his breath, which through the world doth pas,
> Her hollow womb did secretly inspyre,
> And fild her hidden caves with stormie yre
> That she conceiv'd; and trebling the dew time,
> In which the wombes of wemen doe expyre
> Brought forth this monstrous masse of earthly slime,
> Puft up with emptie wynd, and fild with sinfull cryme.[1]

Scholars have often noted, following Heninger, that Orgoglio is by this description a spiritual threat, representing the hot air of the deadly Christian sin of pride as "stormie yre" is "air and ire" in a pun noted by Heninger.[2] The result is a creature that erupts from earth as a Titan but also as a bubble of "emptie wynd" and "sinfull crime." Others have read

Orgoglio in allegorical terms to determine that he is "lust" personified, and still others combine these categories of sin to specify that Orgoglio stands for "carnal pride."[3]

Orgoglio is also, in the terms Spenser offers in the passage, the autonomous, willful progeny of a mother earth. He is nothing less than massive, taking three times as long to gestate as humans. And his lineage mythologically is as long as that through his mother, Earth, and father, Aeolus, human keeper of the wind; in myth, his story dates in written form at least to Hesiod's Titanomachy in *Theogony*. As Maurice Evans observes, Spenser's Titans:

> embody man's fight with the vices and passions of the Fall; and the lore of Gaea, The Earth, forms a link between the classical and the Christian, a prolonged pun uniting two mythologies. Man's vices and rebellious passions spring from his fallen earthly nature just as the monsters and giants who challenge Jove are bred by Gaea.[4]

Just so, the combination of paradigms allows for a range of descriptions that is both pleasurable and complex, helping give full form to the depiction of this giant foe. Evans concludes, "In light of this double background of classical myth and Christian metaphor, the many references to the earth throughout the poem contain a strong and punning irony" (37). I would suggest an alternative reading: Spenser follows Ovid and Aristotle too closely throughout his epic to reduce references to earth into situations of "punning irony." Spenser instead more distinctly draws from a strong rhetorical palate here and throughout the epic to render earth complexly. Moreover, he shows in equal representation the polytheistic, materialist, and Christian accounts for Orgoglio's birth, and, because of this, Orgoglio is yet more entertaining and formidable than has been realized.

Facing this fully loaded triple threat, the largely untested Redcrosse Knight cannot stand steadfast. Orgoglio takes him prisoner and only with the help of England's famed prince Arthur will he escape. Later in the canto, after his escape, Redcrosse then encounters the dragon, which meteorophysiologically is a more powerful adversary, understood as a volcano and thus as an extension of an earthquake in that the volcano has all of the properties of an earthquake but adds to them the spewing of smoke and sulfur. Spenser's dragon is a creature of these attributes, increasing the stakes for Redcrosse who will in facing this threat have another chance to test his developing heroism, what we will see as his staying power. A threat of even greater spiritual and meteorophysiological magnitude than was Orgoglio, the dragon, Spenser shows, is a fearsome form of earth's progeny, who by both size and reverberating noise causes the earth to quake. The dragon, as a volcano, audibly rumbles and shifts the ground before explosively emitting its pestifcrous flames. We first sense this dragon through Una and Redcrosse, as they approach her homeland of Eden: "they heard a roaring hideous sownd,/That all the ayre with terror filled wyde,/And seemd uneath to shake the stedfast ground" (1.11.4.1–3). Such a "roar" in those very words was commonly associated with quakes, recorded by Aristotle. When later employed in the early modern period, that same roar finds association with other embodiments of the explosive release of pressurized air, as with the roars of great beasts and the blasts of trumpets. This noise signals dangerous change, regardless of the matter out of which the motion emanates: beast or earth, thundercloud, gun barrel, or metal instrument. By the "hideous sownd" alone, "all the ayre" is "with terror filled wyde," and Spenser represents the direst conditions ahead—an obvious warning to turn back.

Advancing nevertheless, Una and Redcrosse will encounter this volcano. At the time, as a species of earthquake, it seemed that the material and efficient cause of volcanos were air building within the earth's caverns to cause rumbling so greate as then to cause rock-melting friction. Spenser gives his dragon just these features:

> From his infernall fournace forth he threw
> Huge flames, that dimmed all the hevens light,
> Enrold in duskish smoke and brimstone blew;
> As burning Aetna from his boyling stew
> Doth belch out flames, and rockes in peeces broke,
> And ragged ribs of mountains molten new,
> Enwrapt in coleblacke clowds and filthy smoke,
> That al the land with stench, and heven with horror choke.
> (1.11.44.2–9)

Alone, Redcrosse faces what is in materialist terms a host of heat-producing *meteora*, including "brimstone," which according to Fulke was a "lyquor concreat" in the "midle betwene metals & stones . . . which semeth to be the matter of all drie & whot qualities" (sig.H8r). This meteor "causeth the hot Bathes, and burneth in Aetna, of Scicilia, and Vesuvius of Italye, casting up the Pumise stones" (sig. J1r).[5] Sulfur also causes a powerfully bad odor, and it is this that the volcanic dragon breathes out, spreading a "stench" over "al the land" such that "heven with horror choke."[6] England was never the home of volcanoes, making them a greater source of wonder for English readers. To this point, the last words William Fulke shares in his treatment of "brimstone" mention in the margin "Aetna & Vesuvius" concluding there is "no place here" to treat the subject at more length (sig. J1r). Readers would not need to learn how to recognize real brimstone but they were keen on learning about real, naturally produced wonders that nevertheless seemed the stuff of myth. Such features were a familiar component of classical epic and legend, as this account from Ovid by way of Golding reminds us:

> Mount *Aetna* with his burning ovens of brimstone shall not byde
> Ay fyrye: neyther was it so for ever erst. For whither
> The earth a living creature bée, and that too breathe out hither
> And thither flame, great store of vents it have in sundry places,
> And that it have the powre too shift those vents in divers caces,
> Now damming theis, now opening those, in moving too and fro:
> Or that the whisking wynds restreynd within the earth bylowe,
> Doo beate the stones ageinst the stones, and other kynd of stuffe
> Of fyrye nature, which doo fall on fyre with every puffe:
> Assoone as those same wynds doo cease, the caues shall streight bée cold.
> (209)

The wonder of the volcano is most apparent, its cause to be pondered at length, demanding the "whither" equivocation to account for various interpretations. Ovid offers both the polytheistic "living creature" earth and the materialist "wyn restreynd" options. Spenser's dragon is just such a marvel. It is overwhelming, threatening all things biotic and abiotic with its tremors, scorching flames, and pestilent breath at once, and its origin is unclear.

What distinguishes this dragon from Ovid and Fulke's volcanos is that when it breathes out its calling card of brimstone, which Spenser also calls "sulphure," this dragon-volcano also shows symptoms of excessive choler. Spenser's representation is plain: this dragon breathes out the hottest and driest of bodily humors thought the equivalent of elemental hot and dry fire, as per Fulke's description of the volcano. He breathes out that smoke- and stench-spewing fire because he also is aflame internally with wrath, one of the seven deadly sins. Spenser tells us, "His blazing eyes, like two bright shining shieldes,/Did burne with wrath, and sparkeld living fyre. . . . So flam'd his eyne with rage and rancorous yre" (1.11.14.1–2, 7). Here we encounter again the "yre" that is pestilent air, a full ecological threat generated from within the recesses of the dragon such that its eyes blaze, unable to hide the internal flames. Here too are the eyes that Shakespeare's Henry V wishes his soldiers could acquire when he instructs them to "lend the eye a terrible aspect;/Let pry through the portage of the head/Like the brass cannon" (3.1.10–12)—a description underscoring the resemblances among all hot, dry materials and motions, men, dragons, cannons, comets.

Orgoglio had seemed the proverbial perfect storm, but for Redcrosse, the greatest challenge is this dragon, and it is worth comparing briefly this Saint George with another, the previously mentioned speaker, Shakespeare's Henry V. As we just saw, he appears to be a master teacher in the art of directed passions, successfully instructing his men to embrace the meteorophysiological transformation necessary to become warriors. According to some of the characters who share his story, Henry V also embodied that meteorophysiological condition he recommends. Gloucester, for example, reports at his funeral:

> England ne'er had a king until his time.
> Virtue he had, deserving to command:
> His brandish'd sword did blind men with his beams:
> His arms spread wider than a dragon's wings;
> His sparking eyes, replete with wrathful fire,
> More dazzled and drove back his enemies
> Than mid-day sun fierce bent against their faces.
> What should I say? his deeds exceed all speech:
> He ne'er lift up his hand but conquered.
> (Shakespeare, *Henry V* [1.1.8–16])

In Gloucester's memory, Henry's military might turns the king into the "dragon[ ]," with "sparkling eyes, replete with wrathful fire," not to mention "wings." Ironically in this depiction, Henry V is at once this creature in legend vanquished by Saint George, defender of England, and he is Saint George the vanquisher too. This directed, choler-fueled power is, for Gloucester, among the wonders of Henry V. It would seem Henry had known how to rage in just the right way, to manage within him the full meteorophysiological power of the dragon while leading a nation and without allowing that fire within to destroy him.

Within the action of the play proper as well, Exeter speaks of these qualities when he warns the King of France that Henry V will come to them ready as a meteorophysiological wonder:

> if you hide the crown
> Even in your hearts, there will he rake for it.
> Therefore in fierce tempest is he coming,
> In thunder and in earthquake, like a Jove,
> That if requiring fail, he will compel.
> (Shakespeare, *Henry V* [2.4.97–101]).

Exeter compares Henry, via simile, to Jove, who comes "in fierce tempest . . . In thunder and in earthquake." The anaphoric "in" emphasizes the notion that Henry directs these macrocosmic forces. The "tempest," "thunder," and "earthquake" are the natural extensions of Henry's righteous indignation and of his enraged material body. The suggestion is that he is unstoppable, deadly, almost supernatural, although the comparison offers polytheistic cosmological coloring rather than Christian shades. "Like a Jove," Henry is more powerful than any human or group of humans, well able to "compel." Unlike an earthquake or a tempest, however, and more like Jove, Henry will only "rake" in the hearts of France "if" demands are not met. The implication is that Henry has harnessed meteorophysiological forces, making him an English wonder and the potential scourge of France rather than an unlawful or overly-zealous tyrant. He is the dragon and dragon-slayer of Gloucester's memorial, keeping company with Spenser's Arthur and even more so with his Britomart, the subject of the second half of this chapter.

Spenser's young Saint George, Redcrosse, will not achieve the stature of Shakespeare's incarnation of George in Henry V, not yet, but he will advance in the process of besting the dragon. This is in part due to the nature of the dragon as more formidable of all possible meteorophysiological and Christian threats—the former literally as an active pestiferous volcano and the latter figuratively as hell. Attending also to the dragon as a psychophysiological threat, it is a challenge impossible to augment, because it can create the kind of extreme suffering that Gail Kern Paster describes as "collaps[ing] the overwhelmed self back into its mutely expressive environment."[7] Redcrosse risks becoming the fire he seeks to extinguish, the wrath he wishes to trade for righteous indignation, or perhaps for equanimity. In other terms informed decidedly by a presentist perspective, when Spenser's Redcrosse faces the dragon, he risks all agency and all ecological balance at once.

Redcrosse stands, and he stands steaedfast, but not entirely due to his own meteorophysiological composure. Divine aid, by way of a *deus ex machina*, causes him to stumble into a well of life and later into balm from the tree of life.[8] These substances spiritually and medicinally heal and protect Redcrosse by removing him from and fortifying his body against the "harmefull pestilence" and "scorching heat" of the dragon (1.11.45 and 1.11.50). Redcrosse then with "importune might" kills the dragon, straight in through the mouth with a sword, and the dragon exhales his very "life/That vanish into smoke and cloudes swift" as meteor (1.11.54). The falling of his body, as that of Orgoglio's, causes earth "underneath/ [to] groane." The effect of that falling, Spenser adds via simile is, "as an huge rocky clift/Whose fals foundacioun waves have washt away," causing an impact enough to "dismay" the polytheistic god of earthquakes and of the ocean, "great *Neptune*." The dragon, like Orgoglio, proves to have been an insubstantial meteor that cannot hold its form and lacks staying power, even if initially it appears as a most formidable foe.

Its dead body is a sign of its insubstantiality but also, interestingly, of the power of even such insubstantial bodies to make impressions on the human mind that are substantial indeed. Like that of the dead giant, the dragon's dead body is significant in an examination of Spenser's representation of meteorophysiological wonder. Both bodies after death are still part of the spectacle, among the attendant effects of the experience of disaster—like those literal and literary following the 1580 earthquake treated in chapter 2. Unlike the giant's dead body, the dragon's body retains its size such that "like an heaped mountaine lay" dead (1.11.54.9), and unlike the body of Orgoglio, the dragon's body continues to signify the terror that its living body had elicited, as a crowd forms around it to extend its threat by their words. Eager to view two wonders—that of a true and conquering knight and that of a dragon's dead body—this crowd forms, "Heaped together in rude rabblement":

> But when they came where that dead Dragon lay,
> Stretcht on the ground in monstrous large extent,
> The sight with ydle feare did them dismay,
> Ne durst approch him nigh, to touch, or once assay.
> (1.12.9.6–10)

The "feare" the crowd experiences is notably "ydle," in the Oxford English Dictionary meaning "[w]ithout foundation: baseless, groundless" (s.v. idle, adjective 2c.); it is the product of little more than imagination, spectacle, and free time. This is a fear that nevertheless acts as real fear and leads the group to an overly attentive anxious speculation:

> Some feard, and fledd; some feard and well it faynd;
> One that would wiser seeme, then all the rest,
> Warnd him not touch, for yet perhaps remaynd
> Some lingring life within his hollow brest,
> Or in his wombe might lurke some hidden nest
> Of many Dragonettes, his fruitfull seed;
> Another saide, that in his eyes did rest
> Yet sparckling fyre, and badd thereof take heed;
> Another said, he saw him move his eyes indeed.
> (1.12.10.1–9)

The crowd makes the dragon serve its turns, each speaker enhancing the marvel by projecting upon it his or her own desires: "Some . . . Or . . . Another saide . . . Another said," spinning tales out of nothing, baseless. What the speculations reveal are the kinds of associations the people make with wonders and all things that defy certain explanation but that are within their conjectural grasp. The conversation had by the gathering of people around the dragon, then, represents the kind reported by Thomas Twyne in his account of the 1580 earthquake and the kind associated often with monstrous births, hailstorms, strange cloud formations, and extreme forms of disease.

The crowd in part also fears what we might call the dragon's aftershocks, as they note it might have "lingering life" within "hollow brest" or "womb" waiting again to erupt. "[H]is eyes" possibly contain "[y]et sparkling fyre," that might be turned back into full flame. The account of what the dragon might still do grows further. Imaginations churn over its meaning and associate it with other experiences and concerns; for example, as:

> One mother, when as her foolehardy chyld
> Did come too neare, and with his talants play,
> Halfe dead through feare, her little babe reuyld,
> And to her gossips gan in counsell say;
> How can I tell, but that his talants may
> Yet scratch my sonne, or rend his tender hand?
> So diversly them selues in vaine they fray;
> Whiles some more bold, to measure him nigh stand,
> To prove how many acres he did spred of land.
> (1.12.9.6–11.9)

With the "[on]e mother['s]" counsel, Spenser shows the danger of unexplained preternatural phenomena such as the dragon; wonders will be appropriated toward self-serving ends, "So diversely them selues in vaine they fray." Even those who would advance an approach that we might call materialist or proto-scientific are tempted "to measure him . . . /To prove how many acres he did spread of land"; they too are among those "folke" who "thus flocked" and are left to their devices while the king, queen, and action of the plot move to ceremonial thanksgiving.

Whatever Spenser's dragon was and whatever the earthquake of 1580 had been, people alter their account of facts, sometimes intentionally but always at least unintentionally, with each new recollection and conversation. The mother who uses the fear and lack of knowledge about dragons to stage her own fraudulent authority on the subject in some respects practices a harmless, humorous version of what Katherine Eggert calls "disknowledge" in which to maintain authority people adhere to convenient but fictional certainty on a subject thus to avoid the discomfort in the inconvenient truth of incomplete knowledge.[9] Spenser's dragon, like the phenomenon of an earthquake in early modern England, becomes many things to many people due to its ability to imprint minds with its striking changeability, scope, and apparent threat—all of which that demand interpretation. The effect can be terrifying but it can also be funny, and here, in a decidedly literary effort, Spenser charts some of the terrain covered by Harvey in his earthquake letter, moving from fearful, serious action to laughter. Spenser's humor may not result in Harvey's giggling, but it reinforces the nature of unexplained events, which need not always be fearful ones; they may be openings for exploration, creation, and joy.

*

What had enabled Arthur almost immediately to face without fear and to expose as hot air the meteorophysiological threat of Orgolgio is a correspondent and opposing meteorophysiological force: that embodied by the metal-stone called adamant. Certainly, Arthur is a true Christian knight, less subject to the power of the sin of pride or of lust than others. Still there is more to Arthur's capacity to defeat Orgoglio than this Christian reading reveals, because it is a matter also of Arthur's meteorophysiological superiority. Spenser gives Arthur a rare and undeniable "staying power" exemplified by his shield that is "all of Diamond perfect pure and cleene/It framed was, one massy entire mould,/Hewen out of Adamant rocke with engines keen" (1.7.33.5–7).[10] It is strong, "of Adamant rock," and bright, able to expose all falsehood, all that is puffed up beyond merit (35.3–4). The moment Orgoglio views Arthur's shield, he is "vanisht quite," so that "of that monstrous mas/Was nothing left, but like an emptie blader was" (1.8.24.8–9). The Christian equation at work in this victory is temperance and holiness besting pride. The polytheistic equivalent of the equation is Arthur's Jove-like thunderbolt of a shield that bests anything the earthly Titan can mount. The materialist equation here is slightly different: perfectly mixed, steadfast *meteora*—here the "Diamond . . . of Adamant rock"—outlasts imperfectly mixed *meteora*, such as the suddenly powerful but as soon dissipated earthquakes. In the words of English meteorologist and theologian William Fulke, diamonds, in contrast, are "earthly, *Meteores* . . . called perfectly mixed, because they wil not easely be chaunged and resolved from that forme which they are in, as be stones, metalles and other mineralls" (sig. A2r). They are the *meteora* least susceptible to change and yet with a quality of resilience that distinguishes them from our common usage for the word "Adamant."

An examination of the premodern understanding and usage of the term "adamant" gives us insight into how it is that premoderns imagined the word as well as how they imagined and experienced the natural world itself. Such an examination also sheds light on how they imagined the quality of their loves, of their heroes and leaders, and of their enemies, as they assigned the term adamant to all of them, not reserving it for one group or another. This aspect of an examination of adamant also might prompt us to reconsider our comparatively impoverished and even simplistic rhetoric related to our heroes, our leaders, to our loves, to those who threaten us, and to the natural world itself. First then, a little more etymology: as early as the sixth century B.C., the term "adamant" was used as a noun, naming a metal of high value—the specific word in use being "adamant." It is the metal Aeschylus imagines as the only one able to restrain the Titan Prometheus, for example.[11] Moreover, in its original Greek context adamant took its name from the word αμαστος *adamastos*, meaning untamable. As this word suggests, it was a term conveying far more than just the hardness we associate with it. It also conveys a complex and yet irreducible kind of resilience. The history of adamant as a metal becomes still more complex with its classification as one of the many *meteora* that, as mentioned a moment ago, were the subjects of meteorology as codified by Aristotle in the fourth century B.C.[12] In his explanation of the entire physical universe, Aristotle identified *meteora* as a middle set of things linking those above with those below the moon. Nevertheless, one could only hope to anticipate the development of *meteora*, such as by tracking their relationship to the seasons to make use of them for the purposes of husbandry or sailing, as one could not isolate them for study over time.

The exceptions among *meteora* were those formed within earth and altering in shape and substance slowly over time: the metals and minerals, including adamant. These were the only *meteora* of high use and, therefore, of high exchange value. We might with Jared Diamond say that they fueled civilization.[13] Although Aristotle did not offer a detailed account of each mineral and metal in the way that he had done for other *meteora* from winds to earthquakes and thunder, he explains in a prior work in the physical lectures that all bodies or things in the universe are either simple or compound (in Fulke's words, "perfectly" or "imperfectly mixed") in both substance and motion.[14] More specifically, all sublunary things are compounds made up of the four elements; thus their motion is determined by the lead element(s) of their composition. For example, because, as in human bodies, earth is the leading element, human bodies stay close and even fall to earth, versus a spark that, like the fire that is its dominant element, will rise. *Meteora* were always in such states of change among elements to make it difficult to assign them to one dominant element, which is to say to one distinct sublunary region, to one kind of motion. Minerals and metals, however, more resembled the four elements themselves and the biological compounds of earth's biotic creatures in that they were so much more predictable, with the staying power mentioned above. In book four of the *Meteorology*, Aristotle reinforces this resemblance among animals, vegetables, and minerals by tracking some of the qualities they share, such as features of malleability, durability, and liquefaction.[15] This, then, is part of the premodern understanding of "adamant."

Because metals and minerals were also useful for daily life, people had already for centuries recorded the details of their compositions, behaviors, and locations, and so we have also quite a series of catalogs of minerals and metals that help us further determine the early modern use of "adamant." In the first century, for example, Pliny the Elder, in the *Natural History* (c. 77 AD), cites *adamas* as being the most precious of all sublunary things a human can acquire. Pliny unequivocally states, "The most highly valued of human possessions, let alone gemstones, is the 'adamas,' which for long was known only to kings, and to very few of them."[16] Pliny shows us that by his time already, adamas had become identified as a mineral, no longer as a metal. He goes on to describe six different kinds of the stone found in different countries around the world and concludes:

> All these stones can be tested upon an anvil, and they are so recalcitrant to blows that an iron hammer head may split in two and

> even the anvil itself be unseated. Indeed, the hardness of "adamas" is indescribable, and so too that property whereby it conquers fire and never becomes heated.
>
> (37.15)

As a stone, the quality of hardness is dominant. This is a slight change from the use of adamant to describe a metal that can be forged into a ring and is at the same time strong. Here one form of strength over another has become dominant: hardness over internal resilience. This choice, as Shigehisa Kuriyama brilliantly demonstrates, is one of western might over eastern flexibility—about which I will say a bit more in a moment.[17]

Pliny's knowledge of adamant is nevertheless complex in a new way, because by his time, this stone of which he writes has another associated quality: the powers of attraction and repulsion:

> The "adamas" has so strong an aversion to the magnet that when it is placed close to the iron it prevents the iron from being attracted away from itself. Or again, if the magnet is moved towards the iron and seizes it, the "adamas" snatches the iron and takes it away. "Adamas" prevails also over poisons and renders them powerless, dispels attacks of wild distraction and drives groundless fears from the mind.
>
> (211. 37.15.61)[18]

These additional, and to our understanding fanciful, meanings adhere to the stone and are deemed worthy of repeating, not only in the catalog of Albertus Magnus but also in the romances and drama of early modern England. In *Occult Knowledge, Science, and Gender on the Shakespearean Stage*, for example, Mary Floyd-Wilson speaks to this understanding of adamant, noting that for centuries the adamant was confused with the loadstone or magnet. This confusion, she explains, lent to the descriptive term "adamant" as including the qualities of attraction and repulsion that early modern writers then used to describe experiences of action at a distance, particularly love and the effect of love potions.[19]

By the first century with Pliny, adamas had become another kind of distinct wonder, with qualities layered one upon the other—all in the absence of what we can call a real substance. One additional detail that early modern poets and dramatists do not appropriate but that illustrates the way meaning accretes to an imaginary wonder is the following, also recorded by Pliny. The only thing able to conquer the adamant stone, he explains, is hot goat's blood:

> This indomitable power, in fact, which sets at naught the two most violent agents in Nature, fire, namely, and iron, is made to yield before the blood of a he-goat. The blood, however must be no otherwise than fresh and warm; the stone, too, must be well steeped in it, and then subjected to repeated blows: and even then, it is apt to

> break both anvils and hammers of iron, if they are not of the very finest temper.
>
> (37.15)

Here then is part of the complexity of this meteor, that a biological substance—goat's blood—is only thing able to stand against it. This mystery makes no kind of sense even to Pliny, who therefore needs to attempt explanation:

> To what spirit of research, or to what accident, are we indebted for this discovery? or what conjecture can it have been, that first led man to experiment upon a thing of such extraordinary value as this, and that, too, with the most unclean of all animals? Surely a discovery, such as this, must have been due solely to the munificence of the gods, and we must look for the reason of it in none of the elementary operations of Nature, but wholly in her will.
>
> (37.15)

Pliny can only consider the information to be a revelation, and beyond this he offers no discussion. Among other things, what the association reveals to us is the relatedness of all sublunary materials and motions: blood, stone, flesh, all composed of elements, all of them Aristotelian compounds made of the same materials, governed by the same laws of motion, and behaving in ways that uncannily to us but more naturally to premoderns resembled each other.

By the time of Albertus Magnus in the thirteenth century, people had repeated this odd set of seeming facts about the adamant to the point that Magnus reports in his *Book of Minerals* of the details Pliny noted but with emphasis on the magnetic quality: "magicians say that, bound on the left arm, it is good against enemies and insanity and wild beasts and savage men, and against disputes and quarrels, and against poisons and attacks of phantasms and nightmares."[20] The sixteenth century English translation of the book titled *The Secrets of Albertus Magnus* offers the full entry in terms intended to instruct a reader in the use of stones to achieve certain effects, advising the following on use of adamant:

> If thou wilte overcome thy enemyes [one should] take the stone, whiche is called Adamas, in English speache, a Diamonde, and it is of shyning colour, & very harde, in so muche that it can not be broken, but by the bloud of a gote, & it groweth in Arabia, or in Cypres. And if it be bounden to the lefte side, it is good agaynst enemies, madnes, wyld beastes, venomouse beastes and cruell men, and agaynst chydyng & brawlynge· & agaynst venyme, and invasion of fantasyes, and some call it Diamas.
>
> (sig. C4r)

In this translation that was reprinted more than a dozen times between 1560 and 1723 are the properties of adamant, save that of a direct association with the lodestone. The magnetic quality nevertheless appears here in hint of its attractive and repulsive forces able to repel all dangerous individuals and substances, even dreams.[21] The value of the stone for Magnus is in this relational quality, not its hardness. This is also the quality that Mary Floyd-Wilson examines when she discusses the early modern belief in hidden sympathetic and antipathetic forces that could draw people and things together or pull them apart without there having been a visible connection between them.[22]

Our current understanding of "adamant" is of a human stance of immobility that is negative because it is monolithic and inflexible, and clearly this is an understanding of adamant at odds with that used to represent the subjects of this chapter and the use of the term by early moderns. A brief survey of Spenser's additional uses of adamant and of John Milton's provides ample evidence for this observation. Spenser not only gives Prince Arthur a "well-tempered" shield made of adaman, he also gives Artegall a sword of metal and adamant that the gods used to subdue the Titans and that is gifted to him by Astraea herself, goddess of Justice and proxy for Queen Elizabeth I. Spenser, further, describes the dragon of book one as having such scales as seemed made of adamant, and it is with "chains of adamant" that Guyon binds Acrasia, "For nothing else might keepe her safe and sound."[23] He also applies the term explicitly in the constitution of "character," describing Artegall's as cast of "th'adamantine mould" such that his Amazonian captors will not be able to make "impression" upon him. Similarly, in his seventeenth century epic, *Paradise Lost*, John Milton imagines the forging of "adamantine chains" to bind Satan and the other fallen angels and the use of "gates of burning Adamant" to seal them Titan-like in their prison.[24] The same material he employs as the substance for the shields and body armor of his angels, and, in *Paradise Regained*, his character Satan describes the strength of the Son of God by way of simile:

> And opportunity I here have had
> To try thee, sift thee, and confess have found thee
> Proof against all temptation as a rock
> Of Adamant, and as a Center, firm
> To the utmost of meer man both wise and good,
> Not more.[25]

Adamant is pressed into a wide range of services by Spenser and Milton, and among the characters who know its power are none other than the goddess of Justice, God, and Satan. These are uses of adamant highlighting not the rigidity or hardness *per se* but resolution, the ability to stand "against," to have staying power. Adamant thus comes to represent the greatest of meteorophysiological conditions by which to endure Hamlet's "slings and arrows of outrageous fortune."[26] It is for this reason that Arthur can best Orgoglio, that being steadfast is the more certain strength.

In this examination is also highlighted a "tactic" in the re-mystification of things recommended by Jane Bennett. As mentioned in chapter 1, she advises that we cultivate a certain "naiveté" by "revisit[ing] and becom[ing] temporarily infected by discredited philosophies of nature" (18). Such "discredited philosophies" include that traced previously, leading to centuries worth of belief in adamant as a metal-mineral, flexible-hard, attractive-repulsive nonliving thing that more resembled biological creatures than other *meteora*. The early modern range of meanings for the word "adamant" is a vestige of a pre-scientific era that had not entirely lost its enchantment with heroes and things, love and places; moreover, employed in full by early modern writers like Spenser and Milton, this range of meanings helps us to gain access to the early modern condition of ecosystemic embeddedness that we have largely lost. It is a sense of ourselves as meteorophysiological—as wind, water, earth, and fire, with potential, motion, metamorphosis, and power.

*

Britomart, the Faerie Queen's knight of chastity, is a character Spenser fashions as a guide for those who are not able to settle for disknowledge even as they suffer in uncertainty and for those not quite yet adamantly tempered through life experiences to resist internally quaking. Specifically, whereas Orgoglio and the Dragon are external, meteorophysiological threats to the Faerie Queene's knights, and whereas Arthur has tempered what we can also call his "mettle" over time, Britomart is initially her own greatest threat. Enclosed within her body—a microcosmic earth unto itself—and with no release, the passions that blow up within her, wind-like, alter Britomart meteorophysiologically, threatening rupture and permanent, horrific transformation.[27] As her nurse Glauce explains, just when Britomart needs sleep, she is most shaken:

> Then doth this wicked evill thee infest,
> And riue with thousand throbs thy thrilled brest;
> Like an huge Aetn' of deepe engulfed gryefe,
> Sorrow is heaped in thy hollow chest,
> Whence foorth it breakes in sighes and anguish ryfe
> As smoke and sulphure mingled with confused stryfe.
> (3.2.32.4–9)

By simile, Britomart is earth that undergoes the throes of "Aetn'" with "throbs" and the expulsive heat of the dragon, letting forth "smoke and sulphure" in her "sighes." Her body is exhausting her from the inside, challenging her with these motions and heat. At this point, she is like Redcrosse at the beginning of his quest, entirely untested in the ways of meteorophysiological challenge that make the hero. Like him, she too lacks a diamond shield and the fortitude with which to stand firm. In the narrator's words, her sore also causes "her alabaster brest . . . to pant and quake,/As it an Earth-quake were" (3.2.42.9). Britomart's passion is Titanic in its pressure within her, swelling, sure to erupt and destroy her, body and soul. She suffers as she moves closer and closer to an impending moral, physiological, and meteorological crisis, which in the case of this soon to be knight of chastity appears most likely to manifest itself as sexual impropriety if it is not remedied. There is something decidedly erotic in this trembling and heat that has neither vent nor direction. The giant and dragon had outlets for their heat, directing their aggressive behavior against identified assailants. Anger takes an object. This moving pain that Britomart suffers does not, yet, and it eats away at her own body.

Ever responsive, Britomart's nurse delivers religious, pagan, Galenic, and herbal treatments for what she comes to diagnose as a form of love melancholy—a form that Marion A. Wells associates more specifically with female love-melancholy attributed to "uterine furor."[28] Within Glauce's regimen is also a form of physical therapy that Thomas Elyot in *The Castel of Helthe* calls "vociferation."[29] This is a forced breathing to regulate breath when it grows short—a therapy I treat at greater length in the next chapter. It is worth noting here that many characters in Spenser's poem use this therapy to relieve their passions by speaking forth their troubles and thus unburdening their hearts.[30] The most poignant of those cases is the heart-relief for Una in canto 1. Abandoned by Redcrosse, Una appears a picture of woe as she travels alone to find him.[31] When she meets Arthur, he discerns by empathy that she is keeping secret an internal wound, and he encourages her to share it:

> But woefull Ladie let me you intrete,
> For to unfold the anguish of your hart:
> Mishaps are maistred by aduice discrete,
> And counsell mittigates the greatest smart;
> Found never help, who never would his hurts impart.
> (1.7.40.5–9)

What Arthur describes is a meteorophysiological phenomenon of windy passion that, trapped within the earthen body, will grow and "breed despair" until it is released. Arthur counsels release through confession ("unfold[ing]") of the "hart" to diminish the source of heat and relieve the pressure. More specifically, the "hurts" will be "impart[ed]," which means "to make another a partaker of"—a notion of communally experiencing and bearing together these sublunary pains brought about by change, by loss.[32] There is no other way ("Found never help"), because to persist in hiding "the anguish," is to increase the risks: the trapped air can become fetid and poisonous, eventually erupting and then spreading to others. This is a feature of respiration Shakespeare understood and imagined as a weapon that might be hurled with intent—the subject of chapter 4. Building internal heat risks self-damage first. Many of Spenser's characters are disfigured their unassuaged passions, including Malacasta in the Castle Joyeous, who burns with a lust that has her letting off "sparkes of fire" (3.1.47.7)); and the hag's son, who might have killed his own mother out of rage caused by his festering internal wound ("tears, nor charms, nor herbs, nor counsel might/Asswage the fury, which his entrails teares" [3.7.21.3–4]).

Arthur and Timias also nearly suffer this fate, and Britomart's wound subjects her to exactly this level of disturbance, threatening annihilation or worse, her soul scheduled for eternal burn that would render her physical suffering light. In her own words, she describes the situation:

> hidden hooke with baite I swallowed.
> Sithens it hath infixed faster hold
> Within my bleeding bowells, and so sore
> Now ranckleth in this same fraile fleshly mould,
> That all mine entrails flow with poisonous gore,
> And th'vlcer growth daily more and more;
> Ne can my ronning sore finde remedee
> (3.2.38.9–39.1–6)

Such is the wound within her entrails, lacking obvious remedy and "flow[ing] with poisonous gore" that speaks again of plague and sulfur, of the dragon and of "Aetna" within. The many therapies of the nurse have provided no relief for this unusually potent wound. The "sore" has become rooted, putrefying, not to be dislodged let alone simply exhaled through conversation as either Spenser's Prince Arthur or humanist Thomas Elyot would advise.[33] Worse, Britomart's sore "growth daily more and more."

Finally, Glauce decides to take her to the wizard Merlin, hoping he can help. He knows exactly what is happening and counsels, "let no whit thee dismay . . . For from thy wombe a famous Progenee/Shall spring, out of the auncient Troian blood" (3.3.21.7–22.6). As the woman who will become Britain's great mother, Britomart shakes and burns not because she will birth a prideful, traitorous Titan like Orgoglio or because she houses within her an untrainable dragon of wrath or cupidity, but because she is strong enough to hold these meteorological forces inside her body and channel them into fruitful progeny—progeny that will advance the bloodline of none other than Spenser's patron Queen Elizabeth I. Returning to materialist terms, I am suggesting that Merlin reveals to Britomart the formal and final causes of her eruptive symptoms that, were they strictly natural, would not have such associated causes.

Armed with this knowledge, which is at least as powerful as the armor she acquires, Britomart learns through trial after trial to manage the powerful meteorophysiological forces that threaten earthquake- and volcano-like to alter her from the inside. Internally and externally fortified, Britomart comes to stand steadfast like Arthur, but without quite entirely his adamantine quality. She more resembles mother earth, able to maintain stability through changing seasons and shifting elements. She might be for Helkiah Crooke the exemplification of his Microcosmos: the human body in which radical alteration is as tempered as in universe itself—a perception of the human body as a little world that is self-balancing. But this suggestion does not do justice to Britomart's continuously passionate—pleasurable and painful—experience through the epic.[34]

In the ensuing cantos, Britomart will stand steadfast against all external threats—among them tempests, earthquakes, and volcano-like eruptions: she withstands a tempest to battle Marinel (3.4.8–10); bests Paridel who explodes to challenge her as if he were the earthquake (3.9.15); and she alone passes through the "flaming fire, ymixt with smouldry smoke,/And stinking Sulphure" (3.11.21.6–7) that prevents entry into the House of Busirane, where Amorette, the beloved of Scudamore, is being kept prisoner. It is with Britomart's passing through these flames and her enduring of additional meteorological challenges as she seeks to free Busirane's prisoner that Spenser most concertedly depicts the command that Britomart has of her own body and of the forces that govern all sublunary change.

Before Britomart attempts passage through Busirane's highly effective gate of flame, for example, she sees her traveling companion Scudamore unable to pass through the gate himself, in spite of his passion to do so. This causes Britomart to conjecture aloud that such an effort, to attempt the crossing, would be "Foolhardy, as the Earthes children, which made/ Batteill against the Gods?" (3.11.22.8–9). In this simile, Spenser has Britomart imagine herself as the prideful Titan making an unlawful attempt to seize power over things and situations beyond her rights. But when Scudamore answers in affirmative to Britomart's question and quickly opts for giving up the quest, this spurs Britomart on, and she suggests, "Rather let try extremeties of chaunce" (24.8). She plays the giant, hazarding all against the fiercest of challenges; instead of failing and finding herself in chthonic imprisonment, she crosses through the fire "as a thunder bolt/ Perceth the yielding ayre, and doth displace/The soring clouds into sad showres ymolt" (25.6–8). Britomart's meteorological force is as lightning to cloud; the fire that had threatened is "ymolt," melted away as rain. According to some scholars, she also in this way figuratively enters a fiery hell that she will harrow in Christ-like fashion.[35] What is certain is that she becomes one of the few knights of British literary history who is giant *and* giant-slayer, dragon and dragon-slayer, kin in many ways then to Shakespeare's Henry V.

As if announcing the masque of Cupid that Busirane will soon stage for her, and in keeping with a reading that sees her as Christ harrowing hell, Britomart's entry into the House triggers an earthquake:

> an hideous storme of winde arose,
> With dreadfull thunder and lightning atwixt,
> And an earthquake, as if it straight would lose
> The worlds foundations from his centre fixt;
> A direfull stench of smoke and sulphure mixt
> Ensewd, whose noyaunce fild the fearefull sted
> From the fourth howre of night untill the sixt.
> (3.12.2.1–7)

The combination of effects is indeed marvelous—a meteorologically perfect storm to scare even the stoutest knights from Busirane's door. The spectacle he creates is by matter and motion like to a tempest, dragon, volcano, tidal wave, Titan, earthquake, plague, and what appears the end of the world itself, "as if it straight would lose/The worlds foundations." But Britomart is uniquely fashioned by this point in the narrative, and in response, Spenser tells us:

> the bold *Britonesse* was nought ydred,
> Though much e[n]mou'd, but stedfast still persevered.
> (ll. 8–9)

Britomart is literally "much e[n]mou'd" by the shaking, but the wonder fails to alter her in any other way. This will be exactly her physical and emotional reaction to Busirane's last effort to shake her literally into fear and retreat. Spenser reports that even after yet another quake, "all that did not her dismaied make/. . . But still with stedfast eye and courage stout" she stood ready (37.3–5). Twice Spenser describes Britomart's reaction to the trembling earth as "stedfast"—a word meaning, poignantly in this context, to hold one's ground.[36] Britomart becomes her own foundation while all around her shifts—the still center in the House of Busirane and, arguably, of the epic as a whole, the knight most like Arthur in all of Spenser's epic.

Britomart stands steadfast, not because she has divine assistance with healing waters and balms, as Redcrosse did, and not entirely because, like Arthur, she has protective armor and masculine resolve, although she has something of these things too. She stands steadfast because she has learned to direct the forces of her generative, material, soon-to-be maternal body through faith in the formal and final causes of her meteorological wound as Merlin had disclosed. Her wound will continue to ache and occasionally to throb noticeably as her challenges and passions increase; it will cause her to tremble physically; but it will no longer cause her to tremble in a fear that threatens dissolution. Britomart harnesses the giant and dragon, the earthquake and the volcano, pride and wrath, to serve her destiny as a true knight of chastity and as Britain's mother.

Such generally remains the message of the 1596 expanded version of the epic, as well, with its Britomart tested and found meteorologically triumphant in Isis Church. Our very last sight of Britomart in the epic shows her consumed in an effort "wisely [to] moderate[ ] her owne smart," controlling "womanish" vacillation between "hope of [Artegall's] success" and "woe" over his absence, "Till through his want her woe did more increase," and she finally decides "that the change of aire and place/ Would change her paine, and sorrow somewhate ease" (5.7.445.3–4). She too will seek new adventures, not because she has one assigned to her by the Faerie Queene or by destiny, but because she has already secured a commitment from Artegall. She must wait out his return, potentially, as Eggert suggests, becoming his Penelope and losing her own role as Odysseus, but she does not sit home.[37] She takes up a more active waiting through constant "change of aire and place" that prevents despair and literally alters internal "paine" and "sorrow" even if only to the degree of this "somewhate," which Arthur also endures as he seeks ever to serve his unattainable love, the Faerie Queene herself. The unfinished epic gives voice to the practice of changing "aire and place" to satisfy sustainably one's rankling wounds. One cannot always only stand resiliently against external threats or with steadfast resolve in response to one's passions, even when one is mother of Britain.

Notes

1. Edmund Spenser, *The Faerie Queen*, ed. A.C. Hamilton, 2nd edition, rev. text by Hiroshi Yamashita and Toshiyuki Suzuki (Harlow, UK: Pearson/Longman, 2001; 2007), 1.7.9.1–9. All references to the *Faerie Queene* will be to this edition and cited by book, canto, stanza, and line.
2. On Orgoglio as pride and especially as Catholic hot air, see S.K. Heninger, "The Orgoglio Episode in *The Faerie Queene*," *ELH* 26.2 (1959): 171–187; Anne Lake Prescott, "Why Arguments Over Communion Matter to Allegory: Or, Why Are Catholics Like Orgoglio?" *Reformation* 6 (2002): 163–170; and for a psychosexual reading of Orgoglio's swelling pride, see Kenneth Gross, *Spenserian Poetics: Idolatry, Iconoclasm, and Magic* (Ithaca: Cornell University Press, 1985), 118–128; and Lauren Silberman, *Transforming Desire: Erotic Knowledge in Books III and IV of The Faerie Queene* (Berkeley: University of California Press, 1995), 55–56.
3. For an overview of approaches to the subject of Orgoglio's allegorical meaning, see A.C. Hamilton, et al., *The Spenser Encyclopaedia* (London: Routledge, 1996), 519. On the association between Spenser's giants and those of Genesis, see H.D. Brumble, *Classical Myths and Legends in the Middle Ages and Renaissance: A Dictionary of Allegorical Meanings* (Westport, CT: Greenwood Press, 1998), 139.
4. Maurice Evans, *Spenser's Anatomy of Heroism* (Cambridge, MA: Cambridge University Press, 1970), 37.

5. See also Fulke's description "*Of Whote Bathes*," in *A Goodly Gallerye* (fol. 58v).
6. Interestingly, these are also many of the conditions associated with bubonic plague—a disease thought caused by miasma, the air tainted by standing water, smoke, and other harmful additives. From a materialist perspective, the elements of air and water putrefy if they are unable to circulate naturally—a claim that Arthur Golding supports in his description of earthquakes generally: "the troublednesse of water even in the deepest welles, yeelding moreover an infected and stinking savour" (sig. B3v). S. K. Heninger notes the reference to Aetna in Spenser here but does not, oddly, note that the comparison in to the dragon, concluding only that "Spenser's description of an eruption of Aetna has all the conventional details" (*A Handbook of Renaissance Meteorology:* With Particular Reference to Elizabethan and Jacobean Literature [New York: Greenwood Press, 1968], 1 f 49). This naturally occurring symptom of an earthquake did not accompany the 1580 quake, he insists, but had the quake been natural rather than supernatural, he reasons, the smell would have been apparent. Thomas Lodge, physician and author of *A Treatise of the Plague* (1603) is among many who discuss the relationship between plague and comets in Aristotelian and Galenic terms at once (Thomas Lodge, *The Divel Conjured* [1596-] sig. I2r-I4r). For more on comets and plague, see, for example, Guillaume de Salluste Du Bartas's oft-quoted, *Du Bartas his devine weekes and workes translated: and dedicated to the Kings most excellent Majestie by Josuah Sylvester* (1611), p. 42 ; John Swan, *Speculum mundi Or A glasse representing the face of the world shewing both that it did begin, and must also end:* (1635), p. 101; Thomas Gainsford, *The glory of England, or A true description of many excellent prerogatives and remarkeable blessings, whereby she triumpheth over all the nations of the world* (1618), p. 38. By the late seventeenth century, efforts to explain the relationship between plagues and comets and between plagues and earthquakes continued. See for example John Goad, *Astro-meteorologica, or, Aphorisms and discourses of the bodies cœlestial, their natures and influences discovered from the variety of the alterations of the air . . . and other secrets of nature* (1686), p. 472. On comets in early modern Europe and particularly on their link to plague, see Sara J. Schechner, *Comets, Popular Culture, and the Birth of Modern Cosmology* (Princeton, NJ: Princeton University Press, 1997), 76, 86, 97. On the comet of 1577 as presaging plague, see Thomas Slack, *The Impact of Plague in Tudor and Stuart England* (Oxford: Clarendon Press, 2000), 26–27. A ballad from 1690 mentions in explicit terms the relationship between the 1665 comet and the plague that began that year. See *The English-mans Advice/That all may leave to live in Sin,/and truly Worship God,/Least he in Anger do begin,/to scourge them with his Rod*, available online at the *English Broadside Ballad Archive* (https://ebba.english.ucsb.edu/ballad/20640/image, accessed 3 June 2017).
7. Paster, "The Tragic Subject and Its Passions," *The Cambridge Companion to Shakespearean Tragedy*, ed. Claire McEachern (Cambridge: Cambridge UP, 2013), 156. See also Paster, "Becoming the Landscape: The Ecology of the Passions in the *Legend of Temperance*," in Floyd-Wilson and Sullivan), 137–152.
8. On wells and springs as *meteora*, see Fulke (sig. H2r-H4v).
9. Katherine Eggert, *Disknowledge: Literature, Alchemy, and the End of Humanism in Renaissance England* (Cambridge: Cambridge University Press, 2015).
10. For more on this shield, see D.C. Allen, "Arthur's Diamond Shield in the *Faerie Queene*," *The Journal of English and Germanic Philology* 36.2 (1937): 234–243.
11. Aeschylus, "*Prometheus Bound*," in *Aeschylus II*, Seth G. Benardete and David Grene, translators (Chicago: University of Chicago Press, 1985).

12. Aristotle's fourth century B.C. *Meteorology* was third in line of five of his lectures on the physical world.
13. Jared M. Diamond, *Guns, Germs, and Steel: The Fates of Human Societies* (New York: W.W. Norton & Co. 1997.
14. Aristotle, *On the Heavens*, Ian Mueller, translator (New York and London: Bloomsbury, 2004), 2.269a, 1–15.
15. Aristotle, *Meteorology*, H.D.P. Lee, translator (Cambridge, MA: Harvard University Press, 1952), 4.319, 251.
16. Pliny, *Natural History*, H. Rackham, translator, Loeb Classical Library (Cambridge, MA: Harvard University Press, 1938), 37.15.
17. Shigehisa Kuriyama, *The Expressiveness of the Body and the Divergence of Greek and Chinese Medicine* (New York: Zone Books, 1999).
18. OED s.v. mineral, noun, 5: "A medicine or poison containing inorganic substances."
19. Mary Floyd-Wilson, *Occult Knowledge, Science, and Gender on the Shakespearean Stage* (Cambridge: Cambridge University Press, 2013), 11, 25–26.
20. Albertus Magnus, *Book of Minerals*, Dorothea Wyckoff, translator (Oxford: Clarendon, 1967), 60.
21. We see in various medieval texts some of these treatments of the stone, as in John Trevisa's Translation of Bartholomaeus Anglicus's *De Proprietatibus Rerum* (ME translation, in 1398) "De adamante: Adamans [L Adamas] is a litil stoon of ynde . . . No þing ouercomeþ it, noþer yren noþer fyre . . . it is ybroke wiþ newe hoote blood; Gravers usen þe peces þerof to signe and to þirle preciouse stoones" (196b). In a medieval lapidary in the Douce 291 manuscript (around 1450), we find the following: "Athemaunde [a variant spelling of adamant] is a stone of his name that man may not overcome; when he hath hit on an anfeld of iren and smytith above with a grete hamer of iren, more is empeired the Anfeld & the hamer then is the stone" (196b). In a late English Chaucerian version of *The Romance of The Rose* at 4181: "The stoon was hard, of ademant, Wherof they made the foundement" (http://quod.lib.umich.edu/cgi/m/mec/med-idx?type=id&id=MED445&egs=all&egdisplay=open, accessed 3 June 2017; thanks to Fiona Tolhurst).
22. Floyd-Wilson, *Occult Knowledge, Science, and Gender on the Shakespearean Stage*.
23. From Spenser, Arthur's shield is "Hewen out of Adamant rocke with engines keen," and it is "well tempered" (1.7.33; 5.11).
24. John Milton, *Paradise Lost*, ed. Merritt Y. Hughes (New York: Odyssey Press, 1935), 1.48, 436.
25. John Milton, *Paradise Regained*, ed. Merritt Y. Hughes, *Paradise Regained, the Minor Poems, and Samson Agonistes* (New York: Odyssey Press, 1937) 4.531–535.
26. William Shakespeare, "*Hamlet*," in Ann Thompson and Neil Taylor, editors, *The Arden Shakespeare*, Third Series (London: Arden Shakespeare, 2006), 3.1.57. All references to this play in this volume will be to this edition, cited by act, scene, and line number in a parenthetical citation.
27. On the concept of "mettle" and the four elements and as a means for describing human character, please see initially Mary Floyd-Wilson, *English Ethnicity and Race in Early Modern Drama* (Cambridge, UK: Cambridge University Press, 2003); and Tiffany Jo Werth, "'Degendered': Spenser's Stonie Age of Man," *Spenser Review* 43.2.48 (Fall 2013, www.english.cam.ac.uk/spenseronline/review/volume-43/issue-432/abstracts/degendered-spensers-stonie-age-of-man, accessed 20 July 2017). On the relationship between winds and passions, see Gail Kern Paster, *Humoring the Body: Emotions and the Shakespearean Stage* (Chicago and London: University

of Chicago Press, 2004), 4 and Shigehisa Kuriyama, *The Expressiveness of the Body and the Divergence of Greek and Chinese Medicine* (New York: Zone Books, 1999), 262.

28. Marion A. Wells, *The Secret Wound: Love-Melancholy and Early Modern Romance* (Stanford, CA: Stanford University Press, 2007), 225. For a thorough discussion of the classical origins of the concept of a wound of love, its representation in early modern romance, and on Britomart's wound as caused by a kind of "conception" and leading to a painful figurative pregnancy, see Wells. See also Kristen Abbott Bennet, "Reconceiving Britomart: Spenser's Shift in the Fashioning of Feminine Virtue Between Books 3 and 5 of *The Faerie Queene*," *The AnaChronisT* 14 (2009): 1–14 (http://seas3.elte.hu/anachronist/2009Bennett.pdf, accessed 2 June 2017) and Jonathan Goldberg, "The Mothers in Book III of *The Faerie Queene*," *Texas Studies in Literature and Language* 17.1 (1975): 5–26, especially the idea of Britomart as "crossing" via Merlin, "From Lovesick Virgin to Founding Mother" (18).
29. Thomas Elyot, *The Castle of Helthe* (London, 1541).
30. See Silberman on what she calls "The Wounds of Adonis"—the love-wounds felt by Timias, Britomart, and Marinell, and their resemblance to the wound Adonis receives from the boar, especially as depicted in Ovid (35–48).
31. Among other sources of talk therapy in the epic, see 2.1.46–48; 5.7.19–20; 6.8.18–19.
32. Oxford English Dictionary, s.v. impart, verb, 1a.
33. In "The Second Boke" of *The Castel of Helthe*, for example, Sir Thomas Elyot explains in detail the proper use of each of the non-naturals and concludes with a chapter entitled "Of Vociferation"—a treatment to be used in times of irregular breath and potential overheating: "The chiefe exercise of the brest and instrumentes of the voice is vociferation, whiche is syngynge, redynge, or crienge, whereof is the propertie, that it purgesth naturall heate, and maketh it also subtyll and stable, and maketh the members of the body substanciall and stronge, resisting diseases." In the directions that follow this description, Elyot recommends walking and singing in a low voice, for "By high crieng and loude redinge, are expelled superfuluous humors" (Sir Thomas Elyot, *The castel of helthe gathered, and made by Syr Thomas Elyot knight, out of the chief authors of phisyke; whereby every man may knowe the state of his owne body, the preservation of helthe, and how to instruct well his phisition in sicknes, that he be not deceyved* [1534], 50–51).
34. Helkiah Crooke, *Mikrokosmographia a description of the body of man* (1615). For more on Britomart's "self-definition and self-control," fortified and directed by Merlin's revelation of her destiny, "Britomart, History, and Prophecy" in Isabel G. Mac Caffrey's *Spenser's Allegory: The Anatomy of Imagination* (Princeton, NJ: Princeton University Press, 1976), 291–313; Mary Adelaide Grellner, "Britomart's Quest for Maturity," *SEL: Studies in English Literature, 1500–1900* 8.1 (1968): 35–43; John C. Bean, "Making the Daimonic Personal: Britomart and Love's Assault in The Faerie Queene," *Modern Language Quarterly* 40 (1979): 237–255; related specifically to Chastity and love's "sacred fire," see Lesley W. Brill, *Studies in English Literature, 1500–1900* 11.1 (1971): 16–21. On Britomart's armor as something that changes over time with her, according to Judith Anderson, see "Britomart's Armor in Spenser's *Faerie Queene*: Reopening Cultural Matters of Gender and Figuration," *English Literary Renaissance* 39.1 (2009): 74–96. I see the armor more specifically as evolving with her as part of her mastery of the elements. On Britomart's spear and Spenser's appropriation of Ariosto for its creation, see Selene Scarsi, *Translating Women in Early Modern*

England Gender in the Elizabethan Versions of Boiardo, Ariosto and Tasso, Anglo-Italian Renaissance Studies Series (Farnham and Burlington: Ashgate, 2010), 157–158; and on Britomart as an embodiment of "*discordia concors*," largely symbolized by her armor borrowed of Saxons and Britons, see Michael O'Connell, *Mirror and Veil: The Historical Dimension of Spenser's* Faerie Queene (Chapel Hill: University of North Carolina Press, 1977), 84.

35. On Britomart as a Christ figure who descends into a metaphorical Hell (her rescue of Amoret from the House of Busirane), the earth trembling as a result, see Matthew A. Fike, "Britomart and the Descent into Hell," *ANQ: A Quarterly Journal of Short Articles, Notes, and Reviews* 10.4 (1997): 13–18.
36. Oxford English Dictionary, s.v. steadfast, adj., etymology. It also means, "fixed or secure in position" (A.1.), and my reading of these particular texts is that they lean toward "secure" more than "fixed."
37. Katherine Eggert, *Showing Like a Queen: Female Authority and Literary Experiment in Spenser, Shakespeare, and Milton* (Philadelphia: University of Pennsylvania Press, 2000), 41.

4 Like an Overcharged Gun

In this chapter, I examine the power of the malediction—a curse created by the willful holding of air in the body to the point of necessary, explosive, damaging discharge. The human malediction is the gunshot and the earthquake, the comet and the lion's roar, because it is a sudden meteorophysiological release of compressed air. What makes the curse fascinating is the notion that human agency is a factor in its creation and hurling. To curse with impact, one must fashion oneself for it—a feat not easy to accomplish or to recommend. For this reason, by examining the conditions for these curses, we gain insight into early modern thinking about the degree to which humans might be able to harness and willfully alter the meteorophysiological processes governing them. Always seemingly extraterrestrial in its fiery composition, and potentially supernatural depending on its interpretation, the curse was always also a material product of a sublunary body out of balance by intent. In this chapter, I focus on Queen Margaret, Shakespeare's near master over the curse, which is to say over the elements and their motions.

*

Until recently, scholars examining the early modern curse in English literature have followed David Bevington and, later, Stephen Greenblatt in taking the curse as primarily a matter of words and both their sociopolitical origins, on one end, and the degree to which later those words prove predictive, on the other.[1] With respect to articulation and efficacy of the curse, David Bevington formulated the leading question more than 25 years ago: "Does the actual pronouncing of certain words form a part of the process in which events are fulfilled?"[2] Here I emphasize his "pronouncing of certain words," such that "events are fulfilled"—two portions of the curse's trajectory or "process" most charted by scholars. Greenblatt's focus on the curse as a product of oppressive sociopolitical conditions rounds out our understanding of the components of the curse, its three essential mechanical parts as most regularly identified.[3] These are constituent components, easily discernible for ready scrutiny, in most early modern representations of the cursing "process" of which Bevington speaks.

Scholars have recently identified a subtle—because largely invisible—aspect of the curse that involves more than its strictly human factors of perceived oppression, words, and narrative closure delivered by efficacy. This aspect is, for starters, a speech act and a performance, emanating from and dependent upon a body. John Kerrigan, for example, observes that the curse was (and is) often perceived of as "a dangerous flow of inner stuff outside the order of the body."[4] It is because of this threatening composition of the curse, something more than words, that Keir Elam considers it as always "contaminated and contaminating."[5] For Elam, what he calls "the contagious curse" is evidence of the contagious and material nature of uttered language itself (15). In *Voice in Motion: Staging Gender, Shaping Sound in Early Modern England*, Gina Bloom also examines the curse as word and breath united to create the voice that is political and potentially hazardous because volatile in an always unstable acoustic environment.[6] Bloom examines the efficacy of and anxiety associated with early modern speech acts, each act composed of breath that—she explains, quoting Francis Bacon—is "'engendered of matter' . . . [and] 'tenacious' in its impact on the environment" (89); it is as if the breath hurled did indeed cling to the subjects in its path, "tenacious" in this clinging and able to alter that with which it comes into contact. This is an apt way of thinking about the early modern experience of inner and outer reciprocity, the sharing of substances and motions.

We should accept as a given, then, that the origin of the curse is some form of perceived or real marginalization of the speaker, that audiences judge the success of the curse by the degree to which the specific words chosen and spoken seem to cause, or at least foretell, some specific events that follow, and that recent research into speech acts and performance is essential in charting the full trajectory of the curse. It is time additionally to see the curse as a potent, volatile, breath-based product willfully generated to run counter to the laws of physics and thus to threaten those in its compass once hurled. This portion of the process for early modern cursing is difficult to pinpoint, because it is all but invisible. It occurs within the body after the subject has experienced some form of marginalization and before a speaker-to-be selects specific words for a curse, often before the articulating character has consciously registered the source of his or her ailment, and well before breath becomes voice.

*

When the air is still inside the body, it is engaged in respiration, the most basic of physiological and meteorological processes known to early moderns. Shakespeare shows us in his representation of this process that it is in respiration's regulation—or we should say anti-regulation—that the curse is either conceived or never comes to be, regardless of the external circumstances. A key to health for all early modern bodies alike was to maintain regular respiration so that the heat of the body neither flamed too hot and consumed the body's vital moisture, what Aristotle calls "exhaustion," nor dwindled into a cold nothingness known as "extinction."[7] Alternating cool inhalation with hot exhalation maintained the internal temperature most appropriate to optimal bodily functioning. Although it is an involuntary activity, respiration can be willfully and/or externally disrupted. Just as a change in the sun's intensity can increase evaporation and alter rain patterns, so can oppression fuel a curse by altering the respiration of a person subject to it. As a salient example, Aristotle explains:

> Take as an illustration what occurs when coals are confined in a brazier. If they are kept covered up continuously by the so-called "choker", they are quickly extinguished, but if the lid is in rapid alternation lifted up and put on again they remain glowing for a long time.[8]

Only with regular respiration—inhalation and exhalation to maintain temperature—is there health.

In thinking about respiration, early moderns imagined themselves as having at least one decided advantage over the earth and other animals that shared their meteorophysiological ecosystem. Under the direction of the will—or, as Spenser's Arthur counsels, also with the aids of "reason" and "faith" (1.7.40–41)—one could rectify irregular respiration by intentionally directed exercise. In a chapter of *The Castel of Helthe* (1539), humanist and author of *The Boke Named the Governour* (1531), Thomas Elyot explains the importance of basic respiratory exercises:

> The chiefe exercyse of the brest and instrumentes of the voyce, is vociferacion, whiche is synging, redyng, or crienge, whereof is the propertie, that it purgeth naturall heate, and maketh it also subtyll and stable, and maketh the membres of the body substancyall and stronge, resystynge diseases.[9]

Especially interesting is the Elyot's account of the importance of "synging, redyng, or cryienge, which as the word "vociferate" means is to carry the voice (OED, s.v. vociferate, verb). It is as if full voice is necessary if one wants to respire fully, as is "chiefe exercyse of the brest and . . . voyce." When breath becomes labored due to sudden passion or illness, past the help of common vociferation, there is a more elaborate regimen:

> He that intendeth to attempt this exercise, after that he hath ben at the stoole, and softly rubbed the lower partes, and washed his hands. Lette hym speake with as base a voyce as he can, and walkynge,

> begynne to synge lowder & lowder, but styll in a base voyce, and to take no hede of sweete tunes or armonye. For that nothynge dothe profyte unto helthe of the body, but to inforce hym selfe to synge greatte, for therby moch ayre drawen in by fetching of breath, thrustyth forth the breast and stomacke, and openeth and inlargeth the poores. By high crienge and lowde readynge, are expellyd superfluouse humours.[10]

After evacuation of all extraneous bodily products—in that "he hath ben at the stoole"—the patient, who is practicing self-healing, stirs the heat in the body by "rub"[bing] and "wash"[ing] to encourage the opening of the pores. By "high crienge and lowed readynge," air is taken in and then expelled, both in great quantity, with emphasis on using a "base voyce." The body is essentially unblocked, allowing for the discharge of "superflouse humours" that had been preventing healthy respiration. The body is as earth in a region where the soil is porous, allowing winds to pass in and out freely; this is in contrast to moist damp where the soil can become dense and trap air, leading to earthquakes. If one could manage in these ways the challenges that sometimes came to what was usually an involuntary act, one could be more assured of avoiding the fiery extremes of exhaustion.

*

Subject to the same motions but put to a different purpose, the body might instead become that earthquake, eruptive and—as I will explain at greater length—pestilently contaminating. Pressing respiration into this service is a feat of respiratory management not to be practiced by the average body. This is a truth Shakespeare gives to Queen Margaret long before she teaches Queen Elizabeth to curse in *Richard III*. Margaret recognizes that many bodies are unfit, unwilling, or unable to direct the powers of nature to achieve the long-term goal of cursing. The Duke of Suffolk, Queen Margaret's lover, is an apt example. When King Henry VI banishes him, Margaret vehemently objects, uttering her first curse in the tetralogy. Rather than join in with her, Suffolk attempts to turn her attention towards himself and the sad case of his exile: "Cease, gentle Queen, these execrations,/And let thy Suffolk take his heavy leave."[11] Already overcharged with heat and eager to exhale more of it, Margaret will not be comforted by his efforts to recast her as "gentle." They fan her inner flame, and Shakespeare gives her words in response that match many hurled at Macbeth by his overbearing wife as Margaret berates Suffolk: "Fie, coward woman and soft-hearted wretch!/Hast thou not spirit to curse thine enemies?"[12] By accusing Suffolk of lacking the spirit to curse, Margaret accuses him of lacking vital heat itself. Suffolk's phlegmatic disposition is notable in a play in which most of the other characters are at once aglow with heat—actively embroiled, as they are, in the ongoing War of the Roses that serves as the primary action of Shakespeare's first tetralogy. Suffolk is a character like Elizabeth, discussed later, who helps me to conclude that although gender appears at issue in who it is may successfully catch fire and hurl it, in the end it is not consistently clear that gender is involved, because Margaret, Elizabeth, Richard, and Suffolk each have entirely different outcomes related to their efforts to curse.

Slow to catch fire, Suffolk is "heavy" and potentially unfit for flaming. As he answers Margaret's rebuke, however, he smolders and sets off a spark that is worth tracing through this extended quotation:

> A plague upon them! Wherefore should I curse them?
> Could curses kill, as doth the mandrake's groan,
> I would invent as bitter searching terms,
> As curst, as harsh and horrible to hear,
> Delivered strongly through my fixed teeth,
> With full as many signs of deadly hate,
> As lean-faced Envy in her loathsome cave.
> My tongue should stumble in mine earnest words,
> Mine eyes should sparkle like the beaten flint,
> Mine hair be fixed on end, as one distract;
> Ay, every joint should seem to curse and ban.
> And even now my burdened heart would break
> Should I not curse them. Poison be their drink!
> . . . . .
> All the full terrors in dark-seated hell—
> (3.2.309–328)

Suffolk begins by offering a curse of a different kind: "A plague upon them!" but the question—"Wherefore should I curse them?"—that follows suggests his initial indifference to the act. As if to say, "Why bother?!" he goes on to imagine aloud what might happen "[c]ould curses kill." If they could, he would curse. But then, as he speaks, his body shows him the power of cursing, taking its cue from the imagined condition he posits, as Henry V knew could happen for his men. Rallying his body with "shoulds" and "woulds," Suffolk sets his teeth and his eyes sparkle. He has become a cannon, ready to explode—a metaphor with more than figurative import. Having advanced himself toward respiratory failure, Suffolk experiences such physiological need to curse that he cannot but do so: "my burdened heart would break,/Should I not . . ."[13] Curses may or may not kill his enemies, but any impact on his enemies has become a secondary concern as his body cries out from "every joint" for a violent, vociferous purgation.

Suffolk's turn from passivity to activity, from cold to hot, and from meditation to cursing illustrates that fire in the human body can be augmented and that the act of cursing is at least as dependent upon internal bodily temperature as it is on the words selected in the final articulation. Here it is not the "bitter searching terms" he says he "would invent" but the heat from imagining and feeling thereby into the condition that makes his eyes "sparkle," his "hair fixed on end"—impossible to fake, showing his body truly charged.[14] The curse can be a potent, material weapon. Margaret's move to cut Suffolk off in the middle of his sentence supports this conclusion:

> Enough sweet Suffolk; thou torment'st thyself,
> And these dread curses, like the sun 'gainst glass,
> Or like an overcharged gun, recoil
> And turns the force of them upon thyself.[15]

As Margaret sees it, Suffolk has indeed become the unfired cannon, the "overcharged gun" or "sun" reflected to an infinitely greater brightness in a mirror, blinding the viewer. He suffers from his own condition of now being primed for the curse. A dragon must breathe out its fire, just as Una vociferates as she "imparts" her woe to Arthur, and Britomart directs hers into a quest for her beloved, but Suffolk is constitutionally neither dragon nor knight, and Margaret intervenes. Self-damage might alone be the result for his body, which until now has shown itself more cold and moist (a Fulkian "clammye" suits [fol. 14v.]) than hot and dry.

Such is the nature of building toward a meteorophysiological disgorging, however the eruption cannot so easily be halted. Suffolk's response to Margaret's effort to stop him is aptly one of protest:

> You bade me ban, and will you bid me leave?
> Now, by the ground that I am banished from,
> Well could I curse away a winter's night
> Though standing naked on a mountain top,
> Where biting cold would never let grass grow,
> And think it but a minute spent in sport.
> (3.2.333–38)

Shakespeare provides us with an enormously entertaining image of the "naked" Suffolk cursing from atop a barren mountain as if it were "sport." The serious view of this image shows us a man whose internal temperature rages so hot that the external conditions and consequences matter not. He is compelled now to blow his top like a volcano, suffering all to release his heat. But Margaret knows the peril in which Suffolk places himself. With her final entreaty, she cools Suffolk's internal heat by taking his hand, "dew[ing] it with [her] mournful tears" (3.2.340) and redirecting his passion back toward her as they prepare to part.

The potential for the curse to damage its host was not lost on Margaret. Although she turns Suffolk away from his rage, she fully exploits her own. By the time we see her in act four of *King Richard III*, Shakespeare has shown us that she is Richard's equivalent in the strategic deployment of toxic inwardness.[16] The anger that fueled bloodshed in the *King Henry VI* plays has come to fuel her maledictions. And it serves her intentions, making her arguably the master of the curse in Shakespeare's plays. Not only do her curses appear to be effective but she alone demonstrates knowledge of the precise meteorophysiological mechanism animating them. Others recognize her skill, including the play's current queen, Elizabeth, who eventually turns to her former enemy for help:

> O, thou didst prophesy the time would come
> That I should wish for thee to help me curse
> That bottled spider, that foul bunch-backed toad.
> . . . . .
> O thou, well skilled in curses, stay awhile
> And teach me how to curse mine enemies.[17]

Although Elizabeth had initially paid little heed to Margaret and her cursing, now Elizabeth invokes her aid. Elizabeth also uses the very words Margaret had earlier spoken to her, showing that she indeed had attended with care to Margaret's earlier warning, when Margaret had called Elizabeth "Poor painted queen" (1.3.240). Margaret had made it plain then that she believed Elizabeth was ensnared in the web of "that bottled spider," Richard, and that she would one day have no other recourse but to turn to Margaret "[t]o help thee curse that poisonous bunch-backed toad" (1.3.241, 245).

Elizabeth's precise entreaty succeeds in its aims, and Margaret takes her as pupil, sharing the secret to her successful cursing, which I repeat for closer examination:

> Forbear to sleep the nights; and fast the days.
> Compare dead happiness with living woe.
> Think that thy babes were sweeter than they were,
> And he that slew them fouler than he is.
> Bettering thy loss makes the bad causer worse.
> Revolving this will teach thee how to curse.
> (4.4.118–23)

With this, Margaret recommends an anti-regimen and indeed a form of self-torture. At the time both historically in the action of the play and when it was staged, each of the points in her prescription—do not sleep, fast, embellish and then dwell upon the negative—was thought physically harmful to enact. When Margaret tells Elizabeth to "forbear to sleep the nights," she prescribes what a quick way to increase the body's heat. As Elyot explains:

> The commoditie of moderate slepe, appereth by this, that naturall heate which is occupied about the matter, whereof procedeth nourishment, is comforted in the places of dygestion, and so digestion is made better, or more perfite by slepe, the body fatter, the mynde more quiete and clere, the humours temperate: and moche watche all thynges happen contrarye.[18]

By forbearing to sleep, one remains on constant "watche," thus denying the body its ability to devote its heat to digestion, which was required for "humours temperate."[19] When she calls for fasting, Margaret again recommends a practice certain to tax the body and further deplete its resources. As Elyot explains regarding "forbearynge to receyve any meate or drynke," one should:

> note wel, that by to moche abstinence, the moysture of the body is withdrawen and consequently the body dryeth, and waxeth leane: naturall heate, by withdrawynge of moysture, is to moche incended, and not fyndynge humour to warke in, tourneth his vyolence to the radicall or substanciall moysture of the bodye, and exhaustynge that humour, bringeth the body into a consumption.[20]

Without moisture added to the body by way of food and drink, the "natural heate" itself becomes immoderately or "to moche" inflamed, to the point of "vyolence" against the other naturals of the body. Step by step, Margaret prescribes for what will be a raging inferno of choler—the condition of civil war within the body with all "radicall or substantial moisture" "exhaust[ed]" or "drawn out," as in the materialist meteorological usage. The human body comes into "consumption," wasting because of feeding upon itself.[21]

Shakespeare adds to her recipe. With her advice to imagine one's children "better than they were" and the murderer "fouler than he is," Margaret counsels self-deception of a dangerous kind. Aware of the power of the imagination to harm and heal, Thomas Elyot and other practitioners of the day counseled the very opposite. In his chapter "of dolour or hevinesse of mynd," Elyot comments on managing child-loss. When one's child has died, he recommends one "call to thy remembrance some chylldern (of whome there is no lyttell nomber) whose lyves eyther for uncorrigyble vyces, or infortunate chaunces, have bene more grevouse unto theyr parentes, thanne the deathe of thy chylderne, ought to be unto the[e]."[22] This effort to use the imagination to alter one's emotions is intended to move one to a state of reduced grief and anger. Margaret would instead have Elizabeth use her mind to fool her body into exacerbated negative passions thought to increase heat and reduce bodily moisture—all the better to flame with rage expelled eventually in cursing. The final point of Margaret's prescription is to stew: "Revolving this will teach thee how to curse." A body needs time to mount its hot curse, from fasting and foregoing sleep to "revolving," or meditating, upon exaggerated versions of past harms.[23]

*

In the line immediately following Margaret's prescription, however, Shakespeare focuses our attention on what Elizabeth perceives is missing: "My words are dull," Elizabeth says, "O quicken them with thine."[24] Elizabeth wants the "words" themselves, the specific, repeatable terms that she assumes are necessary for a successful curse. Lacking these details, she has rephrased her original request of Margaret. In reply to the additional query, Margaret shows that she has understood Elizabeth's desire for something more specific: "Thy woes will make them sharp and pierce like mine" (4.4.125). In this answer, Margaret confirms that the specific words of the curse (the "them" in her reply) are not at issue. The key to this early modern curse is what lies beneath the skin and in the middle space between the initial, external experience of being wronged and the articulation of the words Elizabeth had sought from Margaret—words that she will need to find on her own, after her body is ready. Margaret has become Elizabeth's teacher in cursing and her leader in battle, calling her into a state of meteorophysiological readiness as Henry V had done for his troops at Harfleur, encouraging them to act themselves into material change.[25]

After Margaret has shared her knowledge, she exits the stage and the play. Shakespeare leaves the cursing to her pupils. Queen Elizabeth joins with the Duchess of York in a discussion of Margaret's prescription: "Why should calamity be full of words?"[26] the Duchess asks. In Queen Elizabeth's answer is a milder version of the redirect Margaret had employed earlier:

> Windy attorneys to their clients' woes,
> Airy succeeders of intestate joys,
> Poor breathing orators of miseries,
> Let them have scope, though what they will impart
> Help nothing else, yet do they ease the heart.
> (4.4.127–31)

Elizabeth's focus is on the interior spaces of the body, not on the external manifestation of the curse as a set of words. What curses "will impart" to others with respect to their effects might be "nothing," but they "do" relieve the body of the one uttering them. They "ease the heart" by reducing the heat and pressure; acting as a form of vociferation, they can rectify the humors. Therefore, they should be allowed to "have scope" or liberty from the body.[27]

Apparently convinced by the instruction she has received from Elizabeth—and in many ways following her as Elizabeth herself had followed Margaret—the Duchess appears eager to find some relief for herself, concluding,

> If so, then be not tongue-tied. Go with me,
> And in the breath of bitter words let's smother
> My damned son, that thy two sweet sons smother'd.
> *[Trumpet sounds]*
> The trumpet sounds. Be copious in exclaims.
> (4.4.132–35)

She and Elizabeth will not need steel or gunpowder or even particular words so long as they have plenty of "breath" to sustain their "copious . . . exclaims"; the will to execute the violent purging of the breath, and the opportunity to aim it at Richard. The effect will be to turn upon Richard the kind of anger-exacerbated internal heat that he had used when ordering the murders of the princes, and it is as if the "trumpet" cue is reflective of the eruption to the state soon to be brought into being.[28]

Given the manner of Richard's eventual demise, it seems Margaret has done her job, teaching others how to produce an efficacious curse. Not all of her pupils—on stage or in the audience—would have been willing or able to take her direction as Suffolk hoped to do and as his failure demonstrates. There is such variety in human bodies that some will fail to catch fire, and this, ironically, is a truth Shakespeare gives to Elizabeth.[29] The heat-inspired, post-instruction malediction instead arises in Richard's mother, the Duchess of York, not in; Elizabeth who admits in response to the Duchess, "Though far more cause, yet much less spirit to curse/Abides in me. I say amen to her" (4.4.197–98). With respect to the cause of a curse, she had warrant and knowledge enough to bring Richard down by what the Duchess calls "copious exclaim." What renders her unfit is that which lies within—that which in Hamlet's words "passes show" (1.2.85), what she describes as the "spirit" that "abides in" her.

We are left to wonder how much of this "spirit" is enough, what constitutes it, and to what degree might it be willfully controlled. One approach to answering these questions follows and supports Dympna Callaghan's call to attend not only to the body but to "the messy interactions of matter and consciousness."[30] In contrast with Elizabeth's one-time flirtation with the act of cursing, Margaret's long-term respiration management is a performance of what Michael Schoenfeldt calls the "the regulation of desire" that gave men and women of the period a form of control over their bodies and over their "selves."[31] Far from being trapped by a set of beliefs about bodies that might make her subject to forces pneumatic or political, Margaret fashions a stalwart, revolutionary interior by applying her knowledge of physiology and meteorology to her own body and mind. Her body serves her mind's ends and its own at once. She does this neither with Prospero's books nor Friar Lawrence's herbal knowledge but by working within the materialist paradigms that had governed centuries of thinking about sublunary bodies and the effects of their labors. She disgorges a product of her obstructed, hot body, and by doing so, she challenges the Yorkist state.

Michael Schoenfeldt speaks to just these manipulations of the body as forms of rebellion, of the regulation of passions as a practice of agency within an otherwise oppressed state:

> The self becomes for so many of these writers a little kingdom, filled with insurrectionary forces and in continual need of monitoring from within and without. The very resemblance of this internal kingdom to the larger world of political power can make self-discipline an extension of governmental control. But it can also bestow upon the individual an authority whose government contests rather than buttresses that of the terrestrial monarch it resembles. As Milton in particular shows, the kingdom of the self, arising from a carefully cultivated paradise within, can become the site of political resistance.
>
> (39)

Schoenfeldt attends to Milton's representation of the fallen and politically disempowered "self" that can nevertheless preserve internally a space of autonomy, and we can apply this to Margaret's bodily government, used to upset her enemies and change the political landscape. Of course, she has not a "carefully cultivated paradise" within her but rather a hell—an alternative and more hazardous but nevertheless viable "site of political resistance."

*

Shakespeare fashioned Margaret with a will constituted for directing bodily efforts toward radical self and social reengineering, but ultimately, she fares little better than her greatest enemy, Richard III. Both burn and both realize what we hear character after character articulate: in order to curse, one risks one's very heart, which is to hazard death. In *Henry VI Part III*, Shakespeare shows us Richard, then Duke of Gloucester, suffering physically, having learned of his father's murder:

> I cannot weep, for all my body's moisture
> Scarce serves to quench my furnace-burning heart;

Nor can my tongue unload my heart's great burden,
For selfsame wind that should speak withal
Is kindling coals that fires all my breast,
And burns me up with flames that tears would quench.
To weep is to make less the depth of grief:
Tears then for babes; blows and revenge for me![32]

Passion burns within Richard, firing his "heart" and creating a near suffocation that normally "tears would quench" or other comforts would ease, such as giving the heated "wind" vent by "speak[ing]." But unlike Suffolk, who would likely enjoy passing his time in a "lady's chamber" or to "the lascivious pleasing of a lute," Richard refuses purgation in any form.[33] He intentionally holds his heat within him, resolving never to confess its source or otherwise exhale it but to channel it entirely toward revenge. Like Margaret in this degree, he will not shy from the self-torment that accompanies that revenge, suffer his heart what it will.[34]

His plan for revenge grows in step with the pressure building in his "furnace-burning heart," and soon, Richard imagines he might just succeed in sustaining his ecologically unstable interior. Marveling at himself, at his ability to control the forces in his body, he observes:

Why, I can smile, and murder whiles I smile,
And cry "Content!" to that that grieves my heart,
And wet my cheeks with artificial tears,
And frame my face to all occasions.[35]

We know better, and so did Shakespeare's audiences: "Content" can only come with the kind of regular respiration that Richard refuses. The moisture of "artificial tears" cannot pacify his burning and self-consuming body. It is in this performance of a cool exterior that Richard differs with Margaret and Suffolk, who did nothing to hide their meteorophysiological conditions. Their anger was apparent; each joint and sinew, and in Suffolk's case each hair on his head, make plain their internal fires. Richard is all show, refusing to be the same on the outside as on the inside—a state impossible to maintain when what is on the inside is all but literally a comet, a cannon, the belly of a dragon, a breaking heart.

In the first lines of *King Richard III*, Richard reveals, in what has become his most famous speech, that although literal battles have ended, he is not willing to let the sun go down on his anger.[36] More than this, he will incite others toward hatred, spreading his heat to cause a conflagration even within his own family, all the while keeping his own fire burning but hidden. By doing so, he fools other people. He does not, however, alter or escape the basic tenants of meteorophysiology that govern his sublunary body. Like Margaret, he plays the physician, prescribing for his body, utilizing its heat for his own purposes of revenge, but he comes to differ from Margaret by overestimating the degree of control one can possibly have over natural processes as powerful as respiration.[37] From a strictly Aristotelian perspective, heat must rise or otherwise escape from confinement; otherwise, bodies will rupture. No degree of will can entirely change the process, the literal origin of which was thought embedded in the design of a sublunary and, for Christians, postlapsarian cosmos. One could work intentionally within the laws that governed these bodies, finding ways to manage them. Inflexibility on this point becomes a literalist position whether one assumes one has either full or no control over the body.

Richard's rigid adherence to his own prescription reinforces in another register Katharine Eisaman Maus' conclusion: "Richard's downfall results not from tactical inconsistency or loss of nerve, but because contradictions in the way his inwardness is constituted become increasingly oppressive" (51–52). In other words, "The more he struggles to constitute an inwardness by excluding alternative, 'relational' modes of determining identity, the more he finds himself unwillingly entangled in a relational mode" (53). Maus focuses on Richard's social interactions, revealing a dangerous contrast between his "unexpressed interior" and "theatricalized exterior" (2)—the difference between what he is and what he seems to be, inside and outside. In the terms used to describe early modern sublunary bodies, his cool exterior cannot keep a lid on his hot interior, and the longer he attempts to maintain this form of autonomy from the materials and laws of the ecosystem, the more he risks self-ruin. As Ian Frederick Moulton explains, this undoing began the moment Richard proclaimed, "Tears then for babes; blows and revenge for me!"—his humoral imbalance contributing to his social monstrosity.[38] Richard's "social monstrosity" is certainly partially to blame for his ruin; yet it does not take Margaret or Elizabeth, his mother, or Richmond to undo him. Richard ruins himself from the inside out. He exhales in exclamation upon himself, succumbing to the self-cursing that David Bevington sees, rightly but for other reasons, as the primary cause of Richard's demise.

A microcosmic body running out of sync with macrocosmic processes was a danger to itself; it also threatened the entire system. It is in this light that the heated bodies of Margaret and Richard each burn as actants in a Latourian actor network assemblage that is Shakespeare's War of the Roses, a national conflagration. As Deleuze and Guattari explain, "An assemblage is precisely this increase in the dimensions of a multiplicity that necessarily changes in nature as it expands its connections" (8). Every new movement within and across bodies biotic and abiotic alters the whole ("My kingdom for a horse," for a curse, for rain [5.4.7.]). Margaret and Richard, as well as the Lancaster and York families, kindle flames within that become infectious and lead to far more than curses but to a complex alteration of state. In one of the battles, for example, a character called "son" enters the stage carrying a body. When he takes a closer look at the man in his arms, he exclaims, "O God! It is my father's face,/Whom in this conflict I unwares have killed."[39] Soon after, "father" enters with a body in his arms and says, "But let me see: is this our foeman's face?/ Ah, no, no, no, it is mine only son!" (2.5.82–83). This war has blazed a path from the hearts and stomachs of kings and queens to the extremities of their national body.

The York and Lancaster families have created nothing less than an ecological disaster. Their anger is willfully augmented and erupting as the comet and as the earthquake that William Fulke describes in *A goodly gallerye*:

> For thereof cometh it, that in many places where earthquakes have been, great aboundaunce of smoke, flame, & ashes, is cast out, when the aboundaunce of brymstone that is under the grounde, through violent motion is set on fyre, & breaketh forth[ ]. Finally, who knoweth not, what stynking mynerals and other poysonous stuffe doth growe under the earth? wherfor it is no wonder if well water, before an earthquake, be infected, but rather it is to be marveiled, if after an earthquake, there followe not a grevous pestilence, when the whole masse of infection is blowne abroade.[40]

Fulke explains why the earth quakes and why sometimes the plague follows its rupture. The heat trapped in the body not only grows in volume; it becomes poisonous. We can use this reasoning to account for the damage done by the Yorks and Lancasters, who likewise "blow[ ]" their internal heat "abroad[ ]." Richard's and Margaret's stored heat, augmented for purpose, is pestilent in its toxicity, capable of infecting those within its compass. In this mixture of cursing, earthquakes, and plagues, we also have a useful gesture toward an explanation for the many early modern curses that call down a plague on those in range.[41] Here, the plague-cursed houses are not Montague and Capulet but York and Lancaster, and those in the compass of their curses, of their hot bodies, are animals, fields, and waterways as well as humans.

Shakespeare made Margaret wise in some of the ways of self-fashioning: her exit from the play is not as messy as Richard's, audiences experience a certain pleasurable passion in hearing her curse the loathed king, and she teaches us all something about the etiology and symptoms of the early modern curse. But upon final analysis, Shakespeare does nothing to recommend her experience or her practice. In her, he produced a character who exhibits no notable concern regarding the degree to which willfully revolved passions and an overheated body prime the ecosystem for macrocosmic imbalance and prepare the soul for an eternal burn. Margaret already burns through the full tetralogy, and perhaps even past the point of death,[42] joining Richard III in afflicting self-damage at least as great as any heaped on their enemies. As Proverbs 6:27 asks, "Can a man take fyre in his bosome, & his clothes not be burnt?"[43] According to the meteorophysiological laws governing sublunary bodies and to Shakespeare here, the answer is a resounding "No."

Notes

1. Mary Steible, "Jane Shore and the Politics of Cursing," *Studies in English Literature 1500–1900* 43.1 (2003): 1–18; Allison Thorne, "'O, Lawful Let It Be/That I Have Room . . . to Curse Awhile': Voicing the Nation's Conscience in Female Complaint in *Richard III*, *King John*, and *Henry VIII*," in Willy Maley and Margaret Tudeau-Clayton, editors, *This England, That Shakespeare: New Angles on Englishness and the Bard* (Farnham, Surrey, and Burlington, VT: Ashgate, 2010), 105–124. See also Keith Thomas, *Religion and the Decline of Magic* (Oxford: Oxford University Press, 1997), 502–511; and Kenneth Gross, who reads the curse, following John Kerrigan and others, as a reactive mode that, like vituperation and rumor, can serve as a weapon for those lacking political agency, but who also sees the curse as an utterance that is laden with prior and potential meaning, even love. For Gross, the curse is more than political; it is embedded in familial relationships and stories of creation writ large ("King Lear and the Register of the Curse," in *Shakespeare's Noise* [Chicago and London: University of Chicago Press, 2001], 161–192). In Greenblatt's examination of Caliban's cursing and in Mary Steible's and Allison Thorne's accounts of Queen Margaret's maledictions, for example, characters curse as a result of their oppression, at a loss for other weapons of revolt. On cursing in general, see Ashley Montagu, *The Anatomy of Swearing* (New York: Collier Books, 1973), 105. On other forms of execration in Shakespeare, see especially Geoffrey Hughes, *Swearing: A Social History of Foul Language* (New York: Penguin, 1991), and Frances A. Shirley, *Swearing and Perjury in Shakespeare's Plays* (London: George, Allen, and Unwin, 1979). For engaging examinations of cursing and justice in Shakespeare, see the following: Steible, "Jane Shore and the Politics of Cursing"; Sidney C. Burgoyne, "Cardinal Pandulph and the 'Curse of Rome'," *College Literature* 4.3 (1977): 232–240; and William O. Scott, "The Paradox of Timon's Self Cursing," *Shakespeare Quarterly* 35.3 (1984): 290–304. On the biblical basis for Margaret's words in the cursing of Richard at 1.3.214–224, and the degree to which the act effectively "predicts Richard's damnation," see B.S. Lee, "Queen Margaret's Curse on Richard of Gloucester," *Shakespeare in South Africa* [1994]: 15–21, 20).
2. David Bevington, "'Why Should Calamity Be Full of Words?': The Efficacy of Cursing in *Richard III*," *Iowa State Journal of Research* 56.1 (1981): 9–21, 10.
3. Stephen Greenblatt, "Learning to Curse: Aspects of Linguistic Colonialism in the Sixteenth Century," *Learning to Curse: Essays in Early Modern Cultural Studies* (New York: Routledge, 1990), 16–39.
4. John Kerrigan, *Revenge Tragedy: Aeschylus to Armageddon* (Oxford: Clarendon Press and New York: Oxford University Press, 1996), 129, 54.
5. Keir Elam, "'I'll plague thee for that Word': Language, Performance, and Communicable Disease," in Catherine M.S. Alexander, editor, *The Cambridge Shakespeare Library*, Volume II: Shakespeare Criticism (Cambridge: Cambridge University Press, 2003), 316.
6. Gina Bloom, *Voice in Motion: Staging Gender, Shaping Sound in Early Modern England* (Philadelphia: University of Pennsylvania Press, 2007), 35–38. See also Kate E. Brown and Howard I. Kushman, who discuss the early modern malediction as a means for examining coprolalia, "the convulsive

cursing that often accompanies the disorder known as Tourette Syndrome" ("Eruptive Voices: Coprolalia, Malediction, and the Poetics of Cursing," *New Literary History* 32 [2001]: 537–562, 538). The voice of the subject compelled to curse is at once personal and foreign and is always "eruptive rather than expressive" (539), interrupting the present to point to something past, future, or otherwise outside of time (547).

7. See Aristotle, "*On Youth, Old Age, Life and Death, and Respiration*," ed. G.R.T. Ross, in Jonathan Barnes, ed., *The Complete Works of Aristotle: The Revised Oxford Translation* (Princeton, NJ: Princeton University Press, 1984): 1. 748.
8. On this matter of exhalations, respiration, and speech, please also see Katherine Cox on the relationship between "human utterance" as "bodily exhalation" and its relationship to "dew and condensation, products of meteorological exhalation [which are] mechanisms of environmental utterance" Katherine Cox, "The Power of the Air in Milton's Epic Poetry," *SEL: Studies in English Literature, 1500–1900* 56.1 (2016): 149–170.
9. Sir Thomas Elyot, *The Castel of Helthe*, sig. 52v. In the opening sentence of *The haven of health*, in a chapter entitled "What labour is . . .," London physician Thomas Cogan prescribes exercise for the same reason: "The first worde in order of that golden sentence proposed by *Hippocrates*, is labour, which in this place signifieth exercise. . . . Labour then, or exercise, is a vehement moving, the end whereof is alteration of the breath or winde of man" (*The haven of health chiefely gathered for the comfort of students, and consequently of all those that have a care of their health, amplified upon five words of Hippocrates . . . Hereunto is added a preservation from the pestilence, with a short censure of the late sicknes at Oxford* [1584], 1). According to the *OED*, vociferate is a verb derived from the Latin *voci* or *vox*, meaning "voice," and *ferre*, "to carry," as in the definition "to cry out loudly" (*OED* s.v. vociferate, verb, 1), which is much like the definition "a putting forth the voice, a crying out, or exclaiming" from *The New World of English Words* by Edward Philips (1658). For more on vociferation, see Philip Barrough, *The method of phisicke conteyning the causes, signes, and cures of inward diseases in mans body from the head to the foote* (1583), 117, 121; Helkiah Crooke, *Microkosmographia a description of the body of man* (1615), 240; Thomas Tryon, *The good housewife made a doctor, or, Health's choice and sure friend being a plain way of nature's own prescribing to prevent and cure most diseases incident to men, women, and children by diet and kitchin physic only* (1692), 255.
10. Elyot, *The Castel of Helthe*, sig. 53r. According to Stanford Lehmberg,

> Through [Thomas] More Elyot may have known Erasmus, and it is likely that he studied medicine with Thomas Linacre; the preface to Elyot's book *The Castel of Helthe* states that when he was twenty years old 'a worshipfull Phisition, and one of the most renouned at that time in England, perceiving me by nature enclined to knowledge, read unto me the works of *Galen*' and Hippocrates (Elyot, *Castel of Helth*, sig. A4).
>
> (Stanford Lehmberg, "Elyot, Sir Thomas [*c.*1490–1546]," *Oxford Dictionary of National Biography* [Oxford University Press, 2004; online edn., January 2008; www.oxforddnb.com/view/article/8782, accessed 27 January 2013])

As Lehmberg explains, Elyot perhaps most differed from his co-humanist, Linacre, in his desire to place medical knowledge in the vernacular, giving it a wider audience than Linacre did when translating Galen from Greek to Latin. It is for this reason and for the popular appeal of *The castel* that I rely on it here as a text representing the kind of knowledge some members of Shakespeare's audience were likely to possess. For more on the important, popular efforts of men like Cogan and Elyot, see as a starting place Andrew Wear, *Knowledge and Practice in English Medicine, 1550–1680* (Cambridge: Cambridge University Press, 2000), 44 and *passim.*

11. William Shakespeare, "*King Henry VI, Part 2*," in Ronald Knowles, editor, *The Arden Shakespeare*, Third Series (London: Methuen Drama, 1999), 3.2.300–306. Others have attended to the masculinist narratives of tragedy in which men are forbidden to weep, and certainly one reason that King Henry VI is not a significant part of this discussion is because he lacks the bodily constitution and/or will to blaze. On gender in the first tetralogy, see especially Ian Frederick Moulton, '"A Monster Great Deformed': The Unruly Masculinity of Richard III," *Shakespeare Quarterly* 47.3 (1996): 251–268. On women and anger, see Gwynne Kennedy, *Just Anger: Representing Women's Anger in Early Modern England* (Carbondale and Edwardsville: Southern Illinois Press, 2000).
12. Shakespeare, "*King Henry VI, Part 2*," 3.2.207–208.
13. See also Hamlet's expression of his grief at his father's death and mother's hasty marriage, expressed, "break, my heart, for I must hold my tongue" (1.2.159). Arden Third Series editors Thompson and Taylor call this expression a metaphor, reading its significance as a symptom of a meteorophysiological sensation that appears across genres and purposes in the period (179).
14. On horripilation, I note Jesse Lander's work, in part shared during the 2015 New Histories of Embodiment panel at the Shakespeare Association of America Conference (panel director Gail Kern Paster).
15. Shakespeare, "*King Henry VI, Part 2*," 3.2.329–332.
16. This concept of toxic inwardness may be usefully compared to the "toxic discourse" of ecocriticism Charles Whitney discusses in "Dekker's and Middleton's Plague Pamphlets as Environmental Literature" (Rebecca Totaro and Ernest B. Gilman, *Representing the Plague in Early Modern England*, Routledge Studies in Renaissance Literature and Culture [New York: Routledge, 2011], 202).
17. Shakespeare, "*Richard III*," ed. James Siemon, *Arden Shakespeare*, Third Series (London: Methuen Drama, 2009), 4.4.79–81, 116–117. All references to this play will hereafter appear as in-text citation, citing this edition by act, scene, and line.
18. Elyot, *The Castel of Helthe*, sig. 47r.
19. In "Sleep: Theory and Practice in the Late Renaissance," Karl H. Dannenfeldt concludes that the English Renaissance reader had access to "an array of international opinion, fully developed, on the theory and practice of sleep" (*Journal of the History of Medicine and Allied Science* 41.4 [1986]: 420). For more on the Galenic context for understanding sleep in Shakespeare's works, see Rebecca Totaro, "Securing Sleep in *Hamlet*," *Studies in English Literature 1500–1900* 50.2 (Spring 2010): 407–426; Garrett A. Sullivan Jr., "Sleep, Epic, and Romance in *Antony and Cleopatra*," in Sara Munson Deats, editor, *Antony and Cleopatra: New Critical Essays* (New York and London: Routledge, 2005), 259–273; and Garrett A. Sullivan Jr., *Sleep, Romance and Human Embodiment: Vitality From Spenser to Milton* (Cambridge and New York: Cambridge University Press, 2012), especially 32–33 and 51–56 with attention to the negative effects of too much sleep, the still harmful opposite of overwatching treated by Totaro in "Securing Sleep."
20. Elyot, *The Castel of Helthe*, sig. 56r.
21. See the Oxford English Dictionary, s.v. consumption, noun, definition 2a "Originally: abnormality or loss of humours, resulting in wasting (extreme

weight loss) of the body; such wasting; (*obs.*). Later: disease that causes wasting of the body, *spec.* tuberculosis. Now chiefly *hist.*" See also the definition for exhaust, from the word exhauriate, in which the Latin *ex-* out + *haurīre* to draw combines to mean to draw out, to be used up (OED, s.v. exhauriate, verb, etymology).

22. Elyot, sig. 68r. On grief as the "material link" between the lament and the curse, see Bloom, *Voice in Motion*, 91–94.
23. As in his only other use of "revolving"—in the same play, describing "The deep-revolving witty Buckingham" (4.2.46)—revolving here means to meditate upon (OED s.v. revolve, verb, definitions II.8.a.) Another related meaning is "returning to" (OED s.v. revolve, verb, def. I.4).
24. Shakespeare, *Richard III*, 4.4.124.
25. As discussed in the introduction, Henry V tells his men to "imitate the action of the tiger:/Stiffen the sinews, conjure up the blood, . . . set the teeth and stretch the nostril wide,/Hold hard the breath and bend up every spirit/To his full height" (Shakespeare, "*King Henry V*," 3.1.6–17). This notion of performing one's body into a new material constitution has implications for the study of early modern actors and the degree to which they and their audiences were mindful of any material transformations that might take place in the process of acting. To what extent were actors masters of their bodies and wills, managing more regularly if not better their position in the ecosystem by augmenting potent forces in the service of a specific, limited end? Please see Bloom, *Voice in Motion*; and Carolyn Sale, "Eating Air, Feeling Smells: Hamlet's Theory of Performance," in Mary Floyd-Wilson and Garret A. Sullivan, editors, *Renaissance Drama 35: Embodiment and Environment in Early Modern Drama and Performance* (Evanston, IL: Northwestern University Press, 2006), 145–168. Thanks especially to Mary Floyd-Wilson for calling my attention to this passage in this context.
26. Shakespeare, "*Richard III*," 4.4.126.
27. Oxford English Dictionary, s.v. scope, noun, definition 7a. For a gendered reading of the passage above and, later, of Richard's refusal to weep or curse so that he might augment his heat, see Bloom, *Voice in Motion*, as she follows Moulton ("A Monster Great Deformed") to differentiate between Richard and the women of the tetralogy. With respect to Shakespeare's representation of bodies uttering curses, however, I see little reason to draw heavy gender lines.
28. For more on the trumpet, roar, and sounds heralding earthquakes see chapter 6. On the decided shift in meteorophysiological representation to accommodate gunpowder, which begins in the later sixteenth century and only becomes widespread in the seventeenth, there is room for research, particularly in continuation of that begun by Todd A. Borlik and Randall Martin, among others. See Randall Martin, *Shakespeare and Ecology* (Oxford: Oxford University Press, 2015); Todd A. Borlik, *Ecocriticism and Early Modern English Literature: Green Pastures* (New York: Routledge, 2011). See also in the period, and just as a starting place, Ralph Bohun's *A discourse concerning the origine and properties of wind. With an historicall account of hurricanes, and other tempestuous winds. By R. Bohun Fellow of New Coll: in Oxon* (1671).
29. Some bodies cannot—or really should not try to—catch fire. This was a fact not lost on physicians and those offering medical advice, as here Elyot offers a warning before closing the section on vociferation with a general recommendation of the practice:

    > But notwithstanding this exercise is not used alway and of all persons. For they in whome is aboundant of humours corrupted, or be much

diseased with crudite in the stomacke and veines, those doe I counsaile, to abstaine from the exercise of the voice, least much corrupted juyce or vapours, may thereby bee into all the body distributed.

(sig. 53r)

30. Dympna Callaghan, "Body Problems," *Shakespeare Studies* 29 (2002): 69.
31. Michael C. Schoenfeldt, *Bodies and Selves in Early Modern England: Physiology and Inwardness in Spenser, Shakespeare, Herbert, and Milton*, Cambridge Studies in Renaissance Literature and Culture 34 (Cambridge: Cambridge University Press, 1999), 74.
32. Shakespeare, "*King Henry VI Part 3*," in John D. Cox and Eric Rasmussen, editors, *The Arden Shakespeare*, Third Series (London: Methuen Drama, 2001), 2.1.79–86.
33. Shakespeare, "*King Richard III*," 1.1.12–13.
34. On the role of the heart as the furnace of the body, its generator of vital heat, see William W.E. Slights, *The Heart in the Time of Shakespeare* (Cambridge and London: Cambridge University Press, 2008), 20 and passim. On the trope of the breaking heart with respect to humoral theory, see as a starting place Totaro, "Britomart's Meteorological Wound" *Archiv für das Studium der Neueren Sprachen und Literaturen*. 250 (2013), 16–17; Michael Schoenfeldt, "'Give Sorrow Words': Emotional Loss and the Articulation of Temperament in Early Modern England," in Basil Dufallo and Peggy McCracken, editors, *Dead Lovers: Erotic Bonds and the Study of Premodern Europe* (Ann Arbor: University of Michigan Press, 2006), 148–149; and Stephanie Shirilan, *Robert Burton and the Transformative Powers of Melancholy* (Burlington and Aldershot: Ashgate, 2015), 150–151. The studies above largely examine the broken heart as related to love and to grief. In cursing, Richard, Margaret, Suffolk, and others replace grief with rage, exchanging a potentially dulling passion for the one that turns their hearts instead into those overheated furnaces that would run them headlong toward Aristotelian exhaustion (See Aristotle, *On Youth, Old Age, Life and Death, and Respiration*, 748). On Richard's refusal to cry, the heat thus generated, and the context for this display of "unruly masculinity" that pushes his body to its extremes, see Moulton, "A Monster Great Deformed." On the perception of the heart's autonomy as an organ, see Scott Manning Stevens, "Sacred Heart and Secular Brain," in Carla Mazzio and David Hillman, editors, *The Body in Parts: Fantasies of Corporeality in Early Modern Europe* (New York and London: Routledge, 1997), 263–284. On the larger phenomena that Mazzio calls "The pathos of the unsaid," see her *The Inarticulate Renaissance: Language Trouble in an Age of Eloquence* (Philadelphia: University of Pennsylvania Press, 2009), 209.
35. Shakespeare, "*King Henry VI, Part 3*," 3.2.182–185.
36. Ephesians 4:26: "Be angry, but sinne not: let not the sunne go down upon your wrath" (Geneva).
37. As Katharine Eisaman Maus contends, Richard's choice to isolate himself from others will constrain him, pinching him as Caliban was pinched, to bad dreams and howling at spirits. See Katharine Eisaman Maus, *Inwardness and Theater in the English Renaissance* (Chicago and London: University of Chicago Press, 1995), especially the introduction and chapter 2.
38. Moulton, "A Monster Great Deformed," 260.
39. Shakespeare, 3H6 v.60–61.
40. Fulke, *A goodly gallerye*, fol. 23r.
41. We also have an opening for further examination of these relationships with respect to choler, the hot, dry, anger-producing equivalent of the element of fire. See especially Paster, "Becoming the Landscape: The Ecology of the Passions in the *Legend of Temperance*," 137–152.

42. The real queen died a year before Richard became king in June 1483. Thanks to Richelle Munkhoff for reminding me of this.
43. *The Geneva Bible: A Facsimile of the 1560 Edition* (Peabody, MA: Hendrickson Publishers, Inc., 2007). On the spiritual danger of internal heat and as one example of a sermon on this popular passage from Proverbs, the title alone of John Downame's work here is telling: *Spiritual physicke to cure the diseases of the soule, arising from superfluitie of choller, prescribed out of Gods word Wherein the chollericke man may see the dangerousnesse of this disease of the soule unjust anger, the preservatives to keepe him from the infection thereof, and also fit medicines to restore him to health beeing alreadie subject to this raging passion. Profitable for all to use, seeing all are patients in this desease of impatiencie* (1616).

5 These Signs Have Marked Me Extraordinary

In three extended representations of the experience of a literal earthquake in Shakespeare's plays are traces of Queen Margaret's and Britomart's internal fires and externalized symptoms of impending disruption. In these representations of the literal quake experience, however, Shakespeare shows that such states of disruptive change can be registered without fear and sometimes with joy. His earthquakes in these plays show the movement of characters and audiences into states of unequivocal pleasure thus offering a sustained correction to impressions we might have regarding the representation of early modern earthquakes. Gabriel Harvey's salubrious treatment of the earthquake experience in letter form was not after all a one-off, and this gives reason to reconsider the scope of early modern interpretations of metamorphosis generally and of earthquakes specifically in the period.

When an earthquake literally moves Juliet's nurse in the tragedy of *Romeo and Juliet* (1595); Glendower's mother in the history play *I Henry IV* (1596–97); and Thaisa in the romance of *Pericles* (1607–08), it is in each case a polytheistic rumbling related to that which led to Harvey's tee heeing. It is a sign that mother earth is alive and generative and that all terrestrial bodies are fertile even as they are also mortal. Rather than conveying the fear of earthly death or God's wrath, the earthquakes in these plays herald motherhood and are essential to the self-perception of each associated character. They also make more room for laughter in response to epic change. Every trembling of the body is an invitation to throw off trepidations and of and perhaps celebrate, in the words of *Hamlet*, "the heart-ache, and the thousand natural shocks/That flesh is heir to" (3.1.61–62). These are the radical alterations that the postlapsarian maternal body experiences and then comes by way of bodily memory to recount as the first throes of life. Comparatively speaking, the still earth and still body are far more terrifying to imagine, as the tragic end of *Romeo and Juliet* exemplifies. These plays, then, speak to a certain vitality associated with the sublunary system, with change that itself seems animate and generative.[1]

The extended representations of these earthquake experiences also partake in the humorous treatment of disaster that is common in Shakespeare's more frequent and pithier employment of quakes in his writing. In the comedies, Shakespeare uses the phenomenon to enhance his description of something that is rare. For example, in *All's Well That Ends Well*, the Clown extends by earthquake reference his claim that among women it is so very lucky if there is one good one that the more accurate statement would be that one good one is born only "every blazing star or at an earthquake" (1.3.85–6), such a wonder this is. The comparison adds to the comedy in ways very much like that made by Celia in *As You Like It* and Benedick in *Much Ado About Nothing*. Joking about how difficult it is to get Rosalind to understand her point, Celia suggests, "It is a hard matter for friends/to meet; but mountains may be removed with/ earthquakes, and so encounter" (3.2.181–183). Shakespeare inserts an earthquake into the proverb popular at the time "friends may meet, but mountains never greet"; in doing so, he gives Celia the extremely hopeful view that all may be well despite seemingly insurmountable obstacles. In *Much Ado*, Benedick uses a reference to an earthquake similarly, but here to heighten humor: assuring Claudio and Don Pedro that he will never be a horn-mad married man, he swears that if he proves wrong on that score to "look for an earthquake too, then" (1.2.257). In these cases, the reference to an earthquake is a humorous way to exaggerate a claim, serving to create a memorable near-hyperbolic assertion of the kind Jonathan Culler sees as essential in creating the "extravagance of the lyric."[2] This extravagance is the power of poetry itself to move imaginations and bodies, to generate new ideas and behaviors. This, I believe, is also the staying power not of the literal earthquake, which was a fleeting thing, insubstantial by the measure of all paradigms, but of the early modern earthquake once lodged in imagination, memory, and printed texts where it was sustainable and often responsively flexible to suit individual needs.

In the plays treated in the balance of this essay, are mothers whose generative bodies open under threat of earthquake to creative birth. Their stories stand as counternarratives to the majority of printed works at the time that placed together birth and natural disaster in popular print news. In broadsides from the period, the trend in the depiction of meteorophysiological wonder was relentlessly negative and hortatory such that one of many examples of this trend suffices. In the full title of the 1609 pamphlet, the Christian paradigm forms the basis of interpretation for the coincidence of monstrous births, fire, and earthquake: *A true relation of the birth of three monsters in the city of Namen in Flanders: as also Gods judgement upon an unnaturall sister of the poore womans, mother of these obortive children, whose house was consumed with fire from heaven, and her selfe swallowed into the earth. All which happned the 16. of December last. 1608*. The title assures us that radical material disruptions—the punishing thunderbolt causing fire and the swallowing earth, as well as the inconstant female body—are in God's control. The events, moreover, are tragic, good to share but only for their spiritual instruction in an opportunity to practice casuistry, as Julie Crawford has recently asserted.[3]

*

The purpose for the representation of an earthquake in *I Henry IV* stands in decided distinction to that for most popular treatments of such wonders as well as to that for most of the 1580 earthquake pamphlets. Taking cue more from Shakesepare's comedies, the earthquake in this play figures in what is surely one of the funniest of exchanges between larger than life characters in Shakespeare's history plays. The famous Welshman—self-declared Prince of Wales and challenger to King Henry IV—Owen Glendower boasts of his prowess, responding to the statements made about his reputation by his recent ally Henry Percy (Hotspur). Glendower explains why it is that a man might fear when hearing the name of Owen Glendower:

> I cannot blame him. At my nativity
> The front of heaven was full of fiery shapes,
> Of burning cressets; and at my birth
> The frame and huge foundation of the earth
> Shaked like a coward.
>
> (3.1.12–16)

In yet another reminder of the flexibility with which writers employed the polytheistic, materialist, and Christian paradigms, Glendower's description of his birth sets it up for comparison with Christ's, both as a nativity account and as an account of wonder. Glendower's, however, comes not merely with one bright star in the heavens but with many. It also comes with an earthquake, the earth shaking with fear, "like a coward" at the thought of birthing such a polytheistic Titan. We are reminded of William Harvey's ill-trembling (*motus*) versus fearful-trembling (*metus*) earth in chapter 2 and of the Spenserean Titans of chapter 3, but Glendower's intention is entirely in earnest, even as we experience the humorously hyperbolic nature of his claim. Glendower's claim also runs counter to efforts to decipher it strictly through the paradigms, especially in the claim "earth/Shaked like a coward." The cowardly trembling suggests a male rather than female gender association for earth here, as the word "coward" is not often in the plays ascribed to women. Glendower's earth, like Gabriel Harvey's gentle women, has a "*metus*" and not a "*motus*" kind of trembling. Glendower is rather all over the place in his representation of this phenomenon, though he is generally referring to his birth as wondrous, with so many preternatural signs attending it.

However it is to be read, Glendower's intended boast fails to impress Hotspur, who is quick to mount a challenge: "Why, so it would have done," he says of the quake, and "at the same season if your mother's cat had but kittened, though yourself/had never been born" (17–19). In this retort equating Glendower to one of many "kitten[s]" also born at the same time, without wonder, Hotspur entirely denies the supernatural or any wondrous quality of Glendower's birth. Hotspor implies that anyone who believes as Glendower does—that he or she is consequential, to such a godlike degree, within the sublunary system—misunderstands its fundamental matters and motions and deserves scorn or at least teasing. Hotspur thus employs a materialist view of the situation, one following that of William Fulke, treated in prior chapters. Hotspur also does so with delightful humor, diminishing Glendower's scale, power, and rhetoric, and challenging his reputation as a man of action and his command of reality, of its factual interpretation.

Representing Glendower as one of those "kitten[s]" is an unequivocally direct challenge that Glendower takes up right away, insisting, "I say the earth did shake when I was born" (20). Here is reasserts his point, driving it home with insistence rather than with additional explanation, and Hotspur quickly counters, but with a difference. Hotspur offers lines that confirm Shakespeare's comfort with materialist meteorophysiology:

> O, then the earth shook to see the heavens on fire
> And not in fear of your nativity.
> Diseased nature oftentimes breaks forth
> In strange eruption. Oft the teeming earth
> Is with a kind of colic pinch'd and vex'd
> By the imprisoning of unruly wind
> Within her womb, which, for enlargement striving,
> Shakes the old beldam earth and topples down
> Steeples and moss-grown towers. At your birth
> Our grandam earth, having this distemperature,
> In passion shook.
>
> (3.1.24–33)

According to Hotspur, Glendower's birth did not cause the earth to shake for fear; the earth shook because it was ill, having a fit not after all of Harvey's *Terra metus* but of his *Terrae motus*. It is an earthly illness also recalling that asserted by Nausea via Fleming as evidence of the Apocalypse. This shaking, if it occurred, was not apocalyptic, however. It was, by Hotspur's telling, natural, regularly occurring and thus neither consequential nor remarkable. By these standards, it is a shaking primarily on the order of Aristotle's earthquake, as he describes in the *Meteorology*: "We must suppose the action of the wind in the earth to be analogous to the tremors and throbbing caused in us by the force of the wind contained in our bodies."[4] As seismologist R.M.W. Musson observes:

> This is a nice summary of the Aristotelian theory of earthquakes. In this scene, Owen Glendower is being held up to ridicule, and the audience are expected to side with Hotspur. So the message is that earthquakes are a natural phenomenon, and not supernatural portents. (726)

Shakespeare would likely have heard an explanation for earthquakes such as this via William Fulke or the writers of the 1580 earthquake pamphlets, and he clearly gives Hotsur what we could call a rationale or scientific response to counter Glendower's hyperbolic polytheism.

At the same time, both responding to Glendower's quasi-polytheistic claim and following Aristotle, Hotspur brings the human body into the equation: the earth was sick and trembling like a human, just as Aristotle finds "the action of wind in earth" to be "analogous" to that of "wind contained in our bodies." In each case, the literal wind become trapped in an earthen, heavier body, which is not its natural resting place. The natural properties and actions of all bodies appear here in Hotspur's reply: Glendower's birth was as regular as that of any kitten (human and animal alike), and if the earth shook at that time, it was a natural shaking like a body tremor. With the effort to which Hotspur goes to offer a precise reply to Glendower's claims we see Shakespeare's facility with this paradigm, as noted, and his desire to represent in Hotspur an energy that is relentless in all of its pursuits. Hotspur will not be satisfied until Glendower is educated, and he is willing to stay the course, even meeting Glendower part way by employing a trace of the polytheistic meteorophysiology paradigm. The link between the two paradigms is the view of sublunary bodies as analogous, identically composed and moved. In this effort Hotspur both genders and ages earth, who is "the old beldam." Again recalling Fleming's account of earth by way of Nausea from chapter 2, with focus on Fleming's emphasis on earth growing increasingly ill with age, Hotspur's depiction of this earth as Glendower's grandmother makes her anything but youthfully generative—all the better, then, as an effort to tease and anger Glendower.

This discussion of the cause of the nativity is in miniature the battle between Glendower and Hotspur over who is more the man, more the hero, and who, in related form, is the more able interpreter of himself and of nature—these being among the marks of a successful leader. It is an exchange at once entertaining, memorable, and tellingly adjusted by Shakespeare to suit each of the characters involved, each of their perspectives on the human relationship to the matter and motions of the universe. At the same time, with this scholarly reading of Hotspur's reply, we would be remiss not to call out the scatological humor that is perhaps its most potent fuel. As many have observed, and here quoting Musson again, Shakespeare uses "the reference to intestinal gases" to "ridicule[ ] Glendower" (726). As a point of clarification, however, it is worth stating that although Hotspur associates a wind-stuffed earth with the flatulent body, as per Harvey's implication in his letter exchange with Spenser, Aristotle does not make this exact association between the wind expelled in an earthquake and flatulence. Aristotle explains, "We must think of an earthquake as something like the tremor that often runs through the body after passing water as the wind returns inwards from without in one volume" (593). In other words, in this scenario, with urine emitted, air seeks to inhabit its vacated space. The rumbling after urination can be associated with flatulence, but that seems to be a later stage, because his "tremor" comes from an inflow of air rather than only from its being pent up or from its outflow. This puts too fine a point on the matter, to be sure, because the Aristotelian argument makes room for this movement, so to speak, and this gesture allows for the delightful suggestion from Hotspur that for all of his manly boasting, great Owen Glendower is full of entirely harmless, and a bit embarrassing, gas.

Whomever one chooses to support in this rivalry, identical in each character's explanation of the nativity quake are the processes of respiration and generation thought to govern all sublunary bodies. These meteorophysiological processes were understood well enough by audiences to fuel jokes and elicit the complementary exhalation of laughter. Along with the humor, the scene conveys the power of stories in which the appearance of a natural wonder coincides with a time of significance in the life cycle. Any possible quaking occurring with Glendower's birth becomes part of a story of births in general—supernatural, natural, and even those of cats. The earthquake enhances the banter, the humor, and exposes to view the meteorophysiological composition of all generative events—earthquake, flatulence, birth, and jest. It also adds grandeur to Glendower's larger than life character, which despite Hotspur's efforts cannot quite be diminished. Glendower in many ways remains, in Westmoreland's early words to the king, "irregular and wild" (1.1.40), and, in King Henry IV's terms, a "great magician" (1.3.83). The memory of Glendower, as his name, speaks for him beyond his literal presence on stage, and this is in part, because as others have noted, we can read Glendower's rumblings out of Wales, his threatening to upend the state, as putting the action of *1 Henry IV* into motion and as the movements of a Titan, able to threaten the stable English monarchy, even if earth did not rumble when he was born.[5]

*

A still more exuberant reporter of earthquakes in Shakespeare's canon is Juliet's nurse. As many have noted, Shakespeare uses an earthquake—and as some scholars contend, *the* quake of 1580—as a marker of time in

Romeo and Juliet. He also gives place to this earthquake in his narrative where none exists in his primary source, Arthur Brooke's *Tragical History of Romeus and Juliet* (1562).[6] He does so, moreover, in an ironically hilarious scene after Lady Capulet mentions Juliet's age as being "pretty," with respect to her being marriageable, and requests that the nurse confirm Juliet's age. The nurse takes the opportunity to pinpoint Juliet's age exactly, using the earthquake to do so but in ways we postmodern might not entirely expect:

> Faith, I can tell her age unto an hour.
> . . . .
> Even or odd of all days in the year,
> Come Lammas Eve at night shall she be fourteen.
> Susan and she, God rest all Christian souls,
> Were of an age. Well, Susan is with God;
> She was too good for me.
> But, as I said,
> On Lammas Eve at night shall she be fourteen,
> That shall she, marry! I remember it well.
> 'Tis since the earthquake now eleven years,
> And she was weaned, I never shall forget it,
> Of all the days of the year, upon that day.
> For I had then laid wormwood to my dug,
> Sitting in the sun under the dovehouse wall.
> My lord and you were then at Mantua.
> Nay, I do bear a brain. But as I said . . .[7]

The earthquake functions as a marker of chronological time allowing the nurse to do the computation of Juliet's age based on its unusual occurrence: "'Tis since the earthquake now eleven years/And she was weaned." She does not remember the earthquake for the sake of remembering the quake itself. Instead, as it was for Glendower, the quake is something of a hanger for other memories. Unlike Glendower's memory, which was second hand, this is a memory of the nurse's own experience, linked to intimate feelings and thoughts related to her deceased daughter Susan, to the fact that the Capulets were away in Mantua for Lamas Eva, and to Juliet and her weaning. The evidence for the full marking of the scene on her memory, as an impression, comes in the nurse's words, "I never shall forget it," and "Of all the days of the year, upon that day." The portion of the scene the nurse most remembers is not specified. The earthquake is one portion only. It may or may not be the one around which or on which the others are hung. The memory anchor might be Juliet's weaning, which the nurse also remembers as an essential event in Juliet's life and her own—the slow separation of wetnurse and the child she nursed and raised. This, in the nurse's life, is the more important with respect to her identity, which in this case is founded upon a job well done and a role in Juliet's life closer than that of her parents. That the weaning and the earthquake occur together within this life context marks them as a special yet distincly described signs for the nurse. Thus, on the earthquake registers more profoundly and particularly in this story of personal life events than it does within any paradigmatic cosmology.

Extending this memory of Juliet's first movements toward independence with weaning, walking, and speaking coincident with the quake, the nurse recalls her husband, also deceased by the beginning of the play's action:

> Sitting in the sun under the dovehouse wall.
> My lord and you were then at Mantua.
> Nay, I do bear a brain. But as I said,
> When it did taste the wormwood on the nipple
> Of my dug and felt it bitter, pretty fool,
> To see it tetchy and fall out with the dug
> "Shake!" quoth the dovehouse. 'Twas no need, I trow,
> To bid me trudge.
> And since that time it is eleven years;
> For then she could stand high-lone; nay, by th' rood,
> She could have run and waddled all about,
> For even the day before she broke her brow,
> And then my husband—God be with his soul,
> 'A was a merry man—took up the child:
> "Yea," quoth he, "dost thou fall upon thy face?
> Thou wilt fall backward when thou hast more wit,
> Wilt thou not, Jule?" and, by my holidam,
> The pretty wretch left crying and said "Ay."
> To see now how a jest shall come about!
> I warrant, an I should live a thousand years,
> I never should forget it:
>
> (1.3.27–48)

The very sudden "Shake!" disturbs the "dove-house," but the nurse, in recollection of "[s]itting in the sun" near its "wall," is only moved in joy at the thought of the pre-steps Juliet takes into womanhood, marriage, sex, and on to the birth and weaning of her own children. The earthquake is a marker of the natural life cycle, with an emphasis on comedy celebrating fertility rather than on tragedy ending in death. The continuation of the nurse's narration of the content of the memory has little directly to do with the quake, as Juliet's "fall" and the "jest" took place "the day before," but each component of the memory recalls the rest.

Memories in this way are representations of Deleuze and Guattari's assemblages, dependent upon human and non-human actants in constant motion and exchange, bound together momentarily but identifiably, as in this one, this earthquake memory. In this memory of the quake we can identify the features of an assemblage as described in more detail by Deleuze and Guattari: "On the one hand it is a mechanic assemblage of bodies, of actions and passions, an intermingling of bodies reacting to one another; on the other hand it is a collective assemblage of enunciation, of acts and statements, of incorporeal transformations attributed to bodies. Then on a vertical axis, the assemblage has both territorial sides, or reterritorialized sides, which stabilize it, and cutting edges of deterritorialization, which carry it away" (88).[8] The earthquake memory picks up the nurse's husband's "jest," Juliet's response to it, the joy felt in it, the shaking of the dovehouse, the holiday upon them. Each of these things has potential to take the lead in the memory, depending upon the context in which the nurse recalls or retells the memory, which will be often: "I should live a thousand years, I never should forget it." It is also stabilized, a specific memory with parts that plainly do and do not belong, "territorialized" even if in lived experience those lines shift with change among parts.

More mundane in its composing parts than the tale Glendower shares, this quake experience is nevertheless momentous for the nurse, and in the expansiveness of her recollection, the impression the earthquake has made shows itself to be complex and lasting. Its similarity to Glendower's recollection of his mother's story of nativity depends upon the similar relationship in each account between the earthquake memory and the self-fashioning of identity. The maternal figures find the quake as a marker of the special nature of their children, Glendower and Juliet, and it is therefore also a marker of their own special status as mothers to those children. The nurse knows she is so, knows she is an authority in the Capulet household because of this role she plays in Juliet's life. In her performance of memory and of identity, the nurse is present when Juliet's parents are not; her own daughter is still alive and is Juliet's age; her husband is alive and well, part of the system bringing comfort to Juliet; and together the four of them, *sans* Lady and Lord Capulet, make a family. This gives the nurse the strength to assert her importance in the Capulet household, which means in Verona, and to do so against another story she mgiht tell of increasingly isolat in the world, her family members dying and Juliet grows up. It is her sense of herself, forged from events exemplified in her joyously shared earthquake memory that grants her a kind of resilient presence and staying power in the play.[9]

However humorously and joyously explained by the nurse, the earthquake in *Romeo and Juliet* (and arguably for Glendower) also must function as a *memento mori*. It is a token of the death built into the nurse's otherwise happy memory, as she recalls her daughter and husband; it is a token of death likewise built into the play, as early as the prologue. Juliet issues from the "fatal loins" of the Capulets (5)—a prefiguration not entirely remedied or even placed in parentheses by the nurture and/or nursing undertaken by another family. The earthquake in this memory relates to that which, upon Queen Elizabeth I's death, Thomas Dekker will describe as "the alteration of a State,"[10] a wholesale shift in the terms by which we know the world. Such is the case with Juliet's death. When the nurse discovers Juliet dead, her first wish will be to unwish her own life, every single day, even "that day":

> I must needs wake you. Lady, lady, lady!
> Alas, alas, help, help! My lady's dead!
> O weraday, that ever I was born!
> (4.5.13–15)

In crying "weraday, that ever I was born!" the nurse unwishes this and all of her own days, in what is only the opening of her heavy lamentation. Her "Weraday," interchangeable with "welladay" means "woe the day."[11] It is also a statement that recalls Ezekiel's prophetic warning to the Egyptians:

> The worde of the Lord came againe unto me, saying
> Sonne of man, prophecie, and say,
> Thus saith the Lord God, Houle *&* *crye*,
> Wo be unto this day.
> . . . . .
>
> And the sworde shal come upon Egypt, and feare shalbe in Ethiopia, when the slain shal fall in Egypt, when thei shall take away her multitude, and *when* her fundacions shalbe broken downe.[12]

The Geneva Bible's "wo be unto this day" becomes "Woe worth the day," in the King James version, both citing destruction of Egypt's "fundacions" ("foundations" KJV) and both additionally placing God's words in the mouth of Ezekiel, calling the people to "Houle *&* *crye*" ("Howl ye" KJV). This is the greatest expression of lamentation and, in the play, it is uttered not by Juliet's parents but by the nurse, who is certainly more the keeper of the memory of Juliet's life than anyone, including, perhaps, Romeo. Her immediate wish to undo the day and to undo her life is approximated only by Romeo, as he first exclaims upon hearing of her death, "I defy you, stars!" (5.1.24) and as he soon thereafter makes a plan: "Juliet, I will lie with thee tonight" (34). Both he and the nurse experience the "fundacions" of their worlds "broken down," and I submit that we cannot value Romeo's grief over the nurse's given Shakespeare's efforts to fashion her original joys in Juliet. In intense, repetitive extension of the nurse's lamentation that is entirely absent from Shakespeare's source,[13] Shakespeare has her focus on the nature of this particular day:

> O woe, O woeful, woeful, woeful day!
> Most lamentable day, most woeful day

> That ever, ever I did yet behold!
> O day, O day, O day, O hateful day!
> Never was seen so black a day as this.
> O woeful day, O woeful day!
>
> (4.5.49–54)

Repeating "O woe! O woeful, woeful, woeful day! . . . O day, O day, O day," it is as if the nurse is in the process of creating an impression in her memory. For the nurse and for audiences, this day will obscure the vibrant, maternally body-specific earthquake memory of laughter and living loved ones.

By this earthquake memory, however, Shakespeare has shown the nurse as the most loving bearer of Juliet's memory. Any unshakable "statue in pure gold" (5.3.299) to be raised as a memorial of "true and faithful Juliet" (302) will be mute in comparison. With her memory, Shakespeare builds the nurse into a character unlike that in his source, who in that text was "banisht in her age,/Because that from the parents she/dyd hyde the mariage" (sig. 83v). Shakespeare gives the nurse the right to mourn as a mother. The earthquake memory shows why, shows her having earned that right, even if we are also meant to laugh at her excesses. The quake is a sign, then, of the nurse's motherhood, and for this reason it implies sense of earth as mother, rumbling as if in relationship to her human inhabitants as they undergo life-stage alterations—as here with Juliet's weaning and earlier with Glendower's birth. The earthquake in such a reading is not a sign of death, or at least it is not only so; it is a sign of life in its most joyful fecundity.

In the context of the whole play, Shakespeare's representation of the earthquake participates in additional through lines more easily interpreted by way of the materialist and Christian paradigms. Aristotle had posited a relationship between earthquakes and the plague, the former being a sure sign of the latter, because the air trapped in the earth was thought able to grow putrid and pestilent. When it broke out by means of a quake, it could spread its contagious vapors to humans—a subject of discussion in chapter 4. We would not want to extend to a breaking the point notion that the earthquake of the play, although occurring in memory, is a literary sign foretelling the plague, but it suits such alignments made popular in inexpensive print broadsides. As for the Christian reading, it might feel to us a stretch to posit a Providential God behind the fate of Romeo and Juliet as "star-crossed lovers" (Prologue 6) whose deaths might then follow exactly what precedes them in supernatural warnings of earthquake and plague. The earthquake and the plague would be equal signs to the reader that Romeo and Juliet had veered from a path pleasing to God and that these were warnings to avoid acting upon the passion of such love and rather to seek to tame it, redirect it. An audience member conceivably might have interpreted the play this way, although it is finally more compelling to separate a reading of the plague, which prevents the delivery of the crucial information to Romeo, from the earthquake at Juliet's weaning. To separate them allows each to exist and to signify uniquely as the destructive and generative motions of life, some to be feared and after which one mourns but some to be enjoyed and celebrated, even through tragedy: "Wilt thou not, Jule?" Shakespeare did not make them interchangeable; we might follow his representation rather than so render them, simplistically, as more signs of the lovers' fate.

The memory of the quake also underscores the impossibility of determining causality for complex life events, including earthquakes but also love and death. This view of causality is part of Shakespeare's point, the more overtly drawn in comparison with his source in Brooke. This new reading of the play's constructed causality, of its multiple moving parts, also aids in the efforts of those who want to rehabilitate the play by moving it away from its having been for decades type-cast as part of the "cute Shakespeare" oeuvre of teachable-in-high-school plays, like *A Midsummer Night's Dream.*[14] Like his nurse and his literal earthquake in the play, Shakespeare's *Romeo and Juliet* is complex, a story as much about the joys that come of passionate eruption as of the tragic loss that can also follow in its wake. It is for this reason here that I emphasize one additional feature of the nurse's earthquake memory: for all of its polytheistic joy in generation—rites of womanhood, and jest, for examples—the nurse's words bear no over traces to that paradigm. Nor do they contain a materialist view of the quake. Instead, her words include reference to a Christian God: "Susan and she, God rest all Christian souls,/Were of an age. Well, Susan is with God;/She was too good for me" she says of her daughter, and of her husband "God be with his soul/'A was a merry man." Although the nurse expresses only joy in recalling the quake, her self-identity here is with her family and it resides in repeated assurance that daughter and husband are in heaven. This assurance must be admitted as a significant part of the earthquake memory; it may also stand as evidence for the complexity with which writers at the time shifted and blended representational registers and for the degree to which writers did so to satisfy their readers, who interpreted radical meteorophysiological change in creative ways beyond our efforts to categorize them in hindsight.

*

In the strange play that is Shakespeare and Wilkin's *Pericles, Prince of Tyre*, an earthquake also registers as a sign at once of fecundity and of mortality. Having learned of the sins of King Antiochus, and in flight for his life out of Antioch, Pericles lands in Pentapolis and marries its princess, Thaisa. Returning together to Pericles' homeland of Tyre, they are caught in a tempest, and Thaisa undergoes near death during the shipboard delivery of a daughter, Marina.[15] Pericles speaks thus of the delivery of Marina, addressing her directly by way of a blessing:

> Now, mild may be thy life!
> For a more blusterous birth had never babe;
> Quiet and gentle thy conditions, for
> Thou art the rudeliest welcome to this world
> That ever was prince's child. Happy what follows!
> Thou hast as chiding a nativity
> As fire, air, water, earth, and heaven can make
> To herald thee from the womb.
>
> (3.1.27–34)

Marina's "blusterous birth" helps to account for her name and marks her as a wonder. Her nativity is on the order of Glendower's—human but marked decidedly as rare. It is also something like Orgoglio's, which was accompanied by wind that Spenser describes with the same basic word, "blustring." The word is etymologically "blast," "blaze" and "blow," with early associations not only with tempest related winds but with chance, inordinate passions, as in one who blusters.[16] For Marina, however, *all* elements of the sublunary ecosystem—"fire, air, water, earth"—"herald" her coming, as if with trumpets. And she emerges thus, earthquake-like even by sound, "from the womb." Earth here is not trembling in Titan-caused pain or in fear of the marvel. Instead "heaven" too takes part, making for a "nativity" that is good news, with much of the Christian message implied, save that accompanying Marina's "quiet and gentle . . . conditions" is the seeming death of her mother.

In delivery of Marina Thaisa has what Pericles calls "[a] terrible childbed"; unable to find signs of life in her, Pericles concludes that she has died. When the sailors explain to Pericles that he must throw the body of Thaisa overboard, because to carry a dead body aboard will bring them to certain doom, Pericles questions their "superstition" (50) but concedes. He then turns to lamentation. The nativity of his child is the burial of his wife, it seems. Extending the association between this birth and the elements of the sublunary ecosystem that have abused them, he addresses his wife's dead body:

> A terrible childbed hast thou had, my dear,
> No light, no fire. Th'unfriendly elements
> Forgot thee utterly, nor have I time
> To give thee hallow'd to thy grave, but straight
> Must cast thee, scarcely coffined, in the ooze;
> Where, for a monument upon thy bones,
> And aye-remaining lamps, the belching whale

And humming water must o'erwhelm thy corpse,
Lying with simple shells.

(56–64)

In Thaisa's "terrible childbed," she fulfills what Louis Schwartz has recently recounted as the harrowing experience that is Protestant birthing, whereby the mother demonstrates through patience in suffering her election as one of God's elect.[17] And yet Thaisa is denied not only the normal lying in period of recuperation and the full return to the marriage bed; she is denied a proper earthen grave. No "monument" will mark her death save the "humming water" and creatures who swim over her. Just so, her corpse will be "o'whelm[ed]" by the non-human, inarticulate, recalcitrant nature of material "thing-power," as Jane Bennett describes (3). And, interestingly in contrast to Pericles' certainty that the elements "herald" Marina, he says of Thaisa, "Th'unfriendly elements/Forgot thee utterly." He is wrong, of course, because as we know, Thaisa lives. He is also wrong to misread his own expression of what the body of Thaisa would experience were it nonliving: it would be quite embedded in a thriving, interactive assemblage of the sea, immersed in womb-related but also in sensual "humming water" sung to by the strange but meteorophysiologically appropriate "belching whale." Failing to read his own words, or in our own hasty read of them, the elements appear motivated by fickle, uncaring, or at least inscrutable forces.

The elements will instead not only hum for a moment but they will gently lead her casket to shore. Her "scarcely coffined" body will not be "o'erwhelm[ed]." It will not rest "with simple shells." She will be thought of as the treasure of the deep by the men who will find her. Thaisa's casket is disgorged from the sea, the result of what we now see as a tsunami—an event for which Aristotle accounts by the wind's behavior that also animates earthquakes. Aristotle uses the same elements and motions to account for it, and as a relevant side note, researchers have recently shown that the Dover Strait has through history produced waves they call "meteo-tsunamic" in that we know they are not caused by seismic activity but their size and behavior is like tsunami.[18] The casket is discovered on the shore at Ephesus following what the physician Cerimon describes as a "turbulent and stormy night" (3.2.4). His servant agrees, "I have been in many, but such a night as this/Till now I ne'er endured" (5–6). These remarks regarding the wondrous meteorological phenomenon of the tempest are soon followed by an account of an earthquake that rouses some of Cerimon's gentlemen friends earlier than is their usual habit: "our lodgings, standing bleak upon the sea," the first gentleman explains:

Shook as the earth did quake.
The very principals did seem to rend

And all to topple. Pure surprise and fear
Made me to quit the house.

(3.2.14–18)

This is the combination of tidal wave and earthquake about which Aristotle writes, and its force is such that "[t]he very principals" seemed not just to shake but "to rend." "Principles" are most often interpreted as the rafters of a house, here the men's "lodgings" certainly; they may also refer to the elements themselves; either way, they are constituent parts of a material frame that should not "rend" or, by another early modern synonym "topple."[19] Even these men who are used to a range of natural events, their lodging standing "bleak upon the sea," are caught with "Pure surprise and fear." This is a sign that the elements and the full assemblage of the sublunary ecosystem are witness to a marvel. This wondrous event is also in keeping generally with those in Shakespeare's romances, which as a set seem to ask whether a concordance of the elements is related to God, the passions, or whether it is an entirely natural but nevertheless redemptive pattern built into nature itself—or perhaps a combination of these things.

We hear a worded response to such wonder when the First Servant explains to Cerimon and the gentlemen:

Sir, even now
Did the sea toss upon our shore this chest;
'Tis of some wreck.
. . . .
I never saw so huge a billow, sir,
As tossed it upon shore.

(49–54)

In company, with support by way of other witnesses for his claims, the servant explains the nature of the "billow" that delivered the chest. It was "so huge" it exceeds prior experience—no small claim for a person who lives by the sea, who knows when something is a billow or a wave. According to the Oxford English Dictionary, a "billow" was a swelling of the sea produced by a great wind, which suggests its relationships to earthquakes, when that wind is shoved into the earth rather than moving upon the water.[20] This birth by wind and sea delivers no Orgoglio but rather Thaisa, a living woman, and the motion seems intentional; more perhaps than the label of personification explains, "[d]id the sea toss upon our shore this chest" and did "so huge a billow" also "toss[ ] it." The repeated verb is not forceful but its gentleness itself might imply the intent of the element of water, as sea and as billow, to deliver the casket specially ashore, to safety and for recuperation, in spite of what we are told earlier is a "quak[ing]" of earth that brings fear among those not prone to that emotion.[21]

Like the earthquake in Shakespeare's depiction of Glendower and of Juliet's nurse, this one is also original to Shakespeare. In the tale of the *Apollonius*, part of his fourteenth century *Confessio Amantis*, John Gower shares the Pericles story. In his version, the Queen gives birth to Marina on board the ship, surrounded by men and largely unaided. Gower describes the birth itself in three lines: "So that in anguisse and in sorwe/Sche was delivered al be nyhte/And ded in every mannes syhte;/ Bot natheles for al this wo/A maide child was bore tho."[22] In the source, Pericles, called Apollonius, speaks to neither daughter nor wife. Moreover, when later "the see up caste" and "Right as the corps was throwe alonde,/Ther cam walkende upon the stronde/A worthi clerc, a surgien,/ And ek a gret phisicien." Those who discover Thaisa are not struck by any special wonder and are thus more restrained in their curiosity regarding the possibility "there was somewhat in" the coffin. In a 1594 prose version of the story, Lawrence Twine follows Gower in rendering the ocean birth of Marina and the casket's delivery on shore less than remarkable: "the next day morning," Twine writes, for example, "the waves rolled foorth this chest to the land, and cast [it] ashore." Soon thereafter, "a physition whose name was *Cerimon*, who by chaunce walking abroad upon the shore that day with his schollers, found the chest."[23] Shakespeare enhances the descriptions both of Marina's birth and of Thaisa's initial rebirth, adding meteorophysiological shading to increase the pathos of gain and loss.[24]

The tempest-delivery of the baby Marina; tsunami-delivery of Thaisa's body from the sea; the later physic-enabled resuscitation of Thaisa; and her "buri[al] a second time within [Pericles'] arms" (5.3.43–44)—these scenes together underline the mundane and miraculous role that Shakespeare gives to the trembling of earth. The quakes and tempest are not isolated supernatural events produced by a God to nudge his people along on their predestined paths—the stuff of sermons and inexpensive popular printed works. They also are not events strictly aligned with materialist theory to render them as naturally occurring only, fitting a pattern of predictable respiration and digestion *sans* passion. Like the explosive curse experience that Shakespeare represents in his first tetralogy, the earthquake experiences in these plays are fully ecosystemic, fully representative of the earthquake as assemblage. As a set of represented assemblages, this one shows positive rumblings toward procreation rather than relentlessly horrifying ones threatening death. For these reasons, these earthquakes more resemble the ranklings of Britomart's love wound and the mirthful "tee heeing" Harvey gives his gentlewomen than they do Margaret's curses, the dragon's sulfuric emissions, or, in the next chapter, Donne's exasperated cries of doubt.

Shakespeare's plays together, those treated over two chapters, also raise more questions related to gender than there is time to treat in this volume. And what they raise is not as obvious as one might think. Just as Gabriel Harvey does not actually make his gentlewomen ridiculous (though some might think this his intent), just as he instead makes them astute in their observations and queries and as he calls university learning itself into question, leaving his gentlemen decidedly attentive in their lines of inquiry, Shakespeare represents male and female experiences of meteorophysiology that are predictably gender specific from an initial reading of them but are not at all so with a more thorough reading of the words he supplies to represent them. In the plays, for example, male and female characters equally express the relationship between earthquakes and birth. Moreover, whereas Margaret of the tetralogy is best versed in cursing, the play's queen Elizabeth is unable to do so, leaving Richard and, in other plays King Lear and Caliban, closer to Margaret in the representation of a body mounting and hurling a curse. As already suggested, Margaret is also in closer company with Spenser's Arthur, Britomart, and his non-human dragon, as well as with Shakespeare's Henry V, than with the women depicted in this chapter.

With respect to the obviously gendered earth of the polytheistic paradigm, and this paradigm's imaginative gravity that influenced representations of meteorophysiology well past the time that the paradigm was considered credible, I recommend treatments by Mary Garrard and Carol Gilligan of western patriarchy's efforts to suppress the female body and its associated effects of fearful-trembling and the giggle.[25] We see just these efforts of suppression in the plays and the pamphlets, but here as well, the story is more complicated, because as disruptive and durable as the story of mother earth's possible sentience and procreative passions are her Titanic progeny, which are primarily male in depictions; furthermore, not all of these Titans is potent or able to upturn mother. It is not, therefore, possible to create a tidy enough account of gender and meteorophysiology to fit in with the other goals of this volume. Leaving what deservs ample research and discussion for another volume, I end this chapter instead with a reminder that Shakespeare shows us across genders of characters and types of plays (tragedy, history, romance) that the spontaneity, irrepressibility, and changeability of meteorophysiological change is the very stuff of Culler's extravagant lyric, making for wholes far more than the sums of their parts and for stories that speak to us still in new ways over centuries.

Notes

1. One might conceive of a generally balanced earth prior to Adam and Eve's fall, as John Milton did 50 years later—one in which interaction among the elements would be pain free yet generative, allowing for erotic love that did not so painfully rumble within the body or and children born without the fear, trembling, pain, and trauma following God's curse of Eve. On the Fall and the consequent wounding of the earth in *Paradise Lost*, interpreted by Richard DuRocher and Ken Hiltner as a wound resembling Christ's, see Richard

DuRocher, "The Wounded Earth in Paradise Lost," *Studies in Philology* 93.1 (Winter, 1996): 93–115; and Ken Hiltner, *Renaissance Ecology: Imagining Eden in Milton's England* (Pittsburgh, PA: Duquesne University Press, 2008), 48–51. On the extreme conditions of maternal mortality in these years as associated with Eve's punishment versus Adam's, see Louis Schwartz, *Milton and Maternal Mortality* (New York and Cambridge: Cambridge University Press, 2009). See also Genesis 3:16.

2. On the extravagance of the lyric, see Jonathan D. Culler, *Literary Theory: A Very Short Introduction* (Oxford: Oxford University Press, 2000), 77ff.
3. See Alexandra Walsham, *Providence in Early Modern England* (Oxford: Oxford University Press, 1999); Julie Crawford, *Marvelous Protestantism: Monstrous Births in Post-Reformation England* (Baltimore, MD and London: The Johns Hopkins University Press, 2005); and Tessa Watt, *Cheap Print and Popular Piety, 1550–1640* (Cambridge: Cambridge University Press, 1994), 288.
4. Jonathan Barnes, *The Complete Works of Aristotle: The Revised Oxford Translation*, Volume 1 (Princeton, NJ: Princeton University Press, 1984), 1.593.
5. The fear and trembling of England is the concern of King Henry IV, whose words open the play:

   So shaken as we are, so wan with care,
   Find we a time for frighted peace to pant,
   And breathe short-winded accents of new broilsgibb
   To be commenced in strands afar remote.
   (1.1.1–4)

   As David Scott Kastan notes, it is the "we" of England and the king together that are "so shaken" by the civil war following the death of *Richard II* that the king can hardly catch his breath to think about commencing the crusade he had promised at his victory. Westmoreland soon reports of "beastly shameless transformation/By those Welshwomen done" upon the bodies of dead English soldiers (1.1.45–46), naming that "irregular and wild Glendower" as the instigator (40), and this new information convinces the king to abandon his plans for the Holy Land in order to address the internal threats that are otherwise sure to prolong England's worst quaking.
6. On the earthquake of 1580 and the dating of *Romeo and Juliet*, see René Weis, "Introduction," in René Weis, editor, *Arden Shakespeare*, Third Series (London: Methuen, 2012), 36–37; Sidney Thomas, "On the Dating of Shakespeare's Early Plays," *Shakespeare Quarterly* 39.2 (Summer, 1988): 187–194, 191; E.A.J. Honigmann, *Shakespeare's Impact on His Contemporaries* (New York: Macmillan, 1982), 68; Sidney Thomas, "The Earthquake in Romeo and Juliet," *Modern Language Notes* 64.6 (June 1949): 417–419; and Maddalena Pennacchia, "The Stones of Rome. Early Earth Sciences in Julius Caesar and Coriolanus," in Maria Del Sapio Garbero, Nancy Isenberg, and Maddalena Pennacchia, editors, *Questioning Bodies in Shakespeare's Rome* (Rome: V&R unipress, 2010), 313–315.
7. William Shakespeare, *Romeo and Juliet*, in Weis, ed., 1.3.11–16–29. All references to the *Romeo and Juliet* will be to this edition and cited in the text by act, scene, and line.
8. See also Jane Bennett, *Vibrant Matter: A Political Ecology of Things* (Durham: Duke University Press, 2010), viii, xvii, and on memory 10, 23. These memories of the earthquake suggest the nature of an unusual historical event as palimpsestic, in the ways Jonathan Gil Harris explains in *Untimely Matter*

in the Time of Shakespeare (Philadelphia: University of Pennsylvania Press, 2009), 17.

9. On the relationship between memory, remembering, and identity, in which remembering is "social performance," see Garrett Sullivan, *Memory and Forgetting in English Renaissance Drama: Shakespeare, Marlowe, Webster* (Cambridge and New York: Cambridge University Press, 2005), 9–15.
10. Thomas Dekker, *The Wonderful Year* (1603), sig. B2v.
11. OED s.v. wellaway, *int.* and *n.* etymology.
12. In the Geneva Bible, see Ezekiel 30:1–4.
13. Compare in Arthur Brooke's version the discovery of Juliet's death that in fact leaves the nurse speechless and gives extended lament only to Lady Capulet (fol. 68r-v).
14. Please see my forthcoming essay, "In Shakespeare's 'Fair Verona," part of Mary Floyd-Wilson and Darryl Chalk's edited volume, *Contagion and the Shakespearean Stage*, forthcoming in the Palgrave Studies in Literature, Science and Medicine series. Please see also Julia Reinhard Lupton, "Cute Shakespeare," in Julia Reinhard Lupton and Thomas Anderson, editors, *Special Issue of JEMCS: Journal of Early Modern Cultural Studies, JEMCS* 16.3 (Summer 2016): 3–5. I am also referring to Julia Reinhard Lupton's paper delivered as part of the panel, "Cut Him Up in Little Stars: Romeo and Juliet Among the Arts," at the World Shakespeare Congress, 1 August 2016; and to and Sianne Ngai's monograph, *Our Aesthetic Categories: Zany, Cute, Interesting* (Cambridge, MA: Harvard University Press, 2012).
15. With respect to the authorship of this play, all scenes from the chorus of act 3 to the end of the play are attributed to Shakespeare; the first two acts of the play are attributed to George Wilkins, whom the Norton Shakespeare editors call "a freelance playwright" and author of *Miseries of Enforced Marriage* (1606) (2715). See also Suzanne Gossett, "Introduction," in William Shakespeare and Suzanne Gossett, editors, *Pericles* (London: Arden Shakespeare, 2004), 1–37.
16. OED s.v. bluster, verb.
17. Louis Schwartz, *Milton and Maternal Mortality* (Cambridge and New York: Cambridge University Press, 2009).
18. Here is Aristotle on the tsunami:

> The combination of a tidal wave with an earthquake is due to the presence of contrary winds. It occurs when the wind which is shaking the earth does not entirely succeed in driving off the sea which another wind is bringing on, but pushes it back and heaps it up in a great mass in one place. Given this situation it follows that when this wind gives way the whole body of the sea, driven on by the other wind, will burst out and overwhelm the land. This is what happened in Achaea. There a south wind was blowing, but outside a north wind; then there was a calm and the wind entered the earth, and then the tidal wave came on and simultaneously there was an earthquake. This was the more violent as the sea allowed no exit to the wind that had entered the earth, but shut it in. So in their struggle with one another the wind caused the earthquake, and the wave by its settling down the inundation.
>
> (Aristotle's "*Meteorology*," Jonathan Barnes, translator, *The Complete Works of Aristotle: The Revised Oxford Translation* [Princeton, NJ: Princeton University Press, 1984]), 1.595

On the occurrence of the "meteo-tsunami" in the Channel, see K. Simon and Edward A. Bryant, "Meteorological Tsunamis in Southern Britain: An Historical Review," *Geographical Review* 99.2 (2009): 146–163, *JSTOR*, www.

jstor.org/stable/40377378, accessed 24 July 2017. For any initiation into examinations of seventeenth century trade-related encounters with extreme weather, particularly in the east, see first Robert Markley, "Daniel Defoe and the Imagined Ecologies of Patagonia," *Philological Quarterly* 93.3 (2014): 295–216.

19. OED s.v. principals, noun II.9.a, and Orgel's Pelican Shakespeare for "principals," noting them as "rafters" (Stephen Orgel ed., *Shakespeare's Pericles: Prince of Tyre* [New York: Penguin Books, 2001], 9n). And here, the OED for s.v. "bleak," adj, 2.a, citing this very line by Shakespeare to mean "Bare of vegetation; exposed: now often with some mixture of sense 3, wind-swept."
20. OED, s.v. billow, noun, 1.
21. OED, s.v. toss, verb, I.1.a "To throw, pitch, or fling about, here and there, or to and fro: expressing the action of wind or wave, or the light, careless, or disdainful action of a person, on something easily moved." The carelessness implied by the first half of this definition works against my claim, but the repetition of "toss" and "on shore" together support a reading of a gentler delivery.
22. John Gower, *This book is intituled confessio amantis, that is to saye in englysshe the confessyon of the lover maad and compyled by Johan Gower squyer* (1483). All citation of this text from John Gower, "*Confessio Amantis*, Book 8," in R.A. Peck and A. Galloway, editors, *Confessio Amantis*, Volume 1 (University of Rochester, Middle English Text Series, 2006, http://d.lib.rochester.edu/teams/text/peck-gower-confessio-amantis-book-8#apollonius, accessed 15 July 2017).
23. Lawrence Twine, *The patterne of painefull adventures* (1594), sig. G4v. So too even does George Wilkins, Shakespeare's collaborator, leave out the earthquake and other extended treatments of meteorology in these scenes in his prose version of the story, *The Painfull Adventures of Pericles Prince of Tyre* (1608), which Suzanne Gossett concludes was most likely written and published *after* the play had been written but before it had been performed, while theaters were closed due to plague (60–61).
24. Here is an expression of Jane Bennett's hope that her work might incite a "tun[ing] in to the strange logic of turbulence" (xi).
25. Carol Gilligan, *The Birth of Pleasure: A New Map of Love* (New York: Vintage Books, 2003); and Mary D. Garrard, *Brunelleschi's Egg: Nature, Art, and Gender in Renaissance Italy* (Berkeley, CA: University of California Press, 2010).

6 These Earthquakes in Himself

Early in the seventeenth century, Thomas Dekker was among many who provided in writing complex and specific examples of Hamlet's "heart-ache, and the thousand natural shocks/That flesh is heir to"—nearly the opposite of some of the scenes of mirth Shakespeare staged in the plays treated in the prior chapter (3.1.61–62). In this chapter, in an initial return to the association of earthquakes with terrifying change, we find in Dekker's *The Wonderful Year* (1603), that these shocks occur as part of England's reaction to the death, after a 45-year reign, of Queen Elizabeth I:

> her departure was so sudden and so strange, that men knew not how to weepe, because they had never bin taught to shed teares of that making. They that durst not speake their sorrowes, whisperd them: they that durst not whisper, sent them foorth in sighes. Oh what an Earth-quake is the alteration of a State! Looke from the Chamber of Presence, to the Farmers cottage, and you shall finde nothing but distraction: the whole Kingdome seems a wildernes, and the people in it are transformed to wild men. The Map of a Countrey so pitifullie distracted by the horor of a change, if you desire perfectlie to behold, cast your eyes then on this that followes.[1]

The "sudden and so strange" change in England's circumstances leaves the people unable to "weepe." They find it difficult to acquire appropriate passage for the enormous burden of their grief. It is as if their passion is so great it cannot be "impart[ed]" in the ways Spenser's Arthur advises Una; nor can it be vociferated, as per Thomas Elyot's advice, discussed in chapters 2 and 4. The English people do what they can under "distract[ion]" to "whisper[ ]" and "sighe[ ]"—exhalations of air insufficiently relieving. Their emotional waters and winds seek outlet, but the shock of Elizabeth I's death seems to close the collective meteorophysiological pores. This condition makes the people, their very bodies, ripe for rupture or for one of its emotional equivalents, such as madness. Unable to respire normally, they resemble the materialist earth ready for a quake and both the erotically enthralled body of Britomart of chapter 3 and the intentionally charged body of Queen Margaret treated in chapter 4—each of which likewise suffers internal pressure that builds and threatens to lead to the "alteration of a State." In Dekker's previously mentioned rendering, that threatened change is the dissolution of culture itself, a return to "wildernes" populated by "wild men."

What England faces with this change in monarch is the threat of nothing less than of Titans, of the everpresent potential for disruption, of change itself. However much institutions and individuals might like and need to imagine it otherwise, this condition of threat is also that of the sublunary experience writ large, which is always a state of change not of stasis. To have imagined Elizabeth I queen forever, as she had seemed to be, is to have dreamt an impossible dream that is exposed harshly with her death. To awaken from it, is to be recalled to instability, to the nation and to one's own body as meteoric, as shifting event not stable thing. This meteoric body, like the "vitality of matter" itself, is (as cited in the Introduction to this volume and according to Bennett) "hard to discern" and then "hard to keep focused on" for it is "as much wind as thing, impetus as entity, a movement always on the way to becoming otherwise, an effluence that is vital and engaged in trajectories but not necessarily intentions." It is a body and a material reality humans are so eager to control. The Queen's death exposes the cracks in the paving over of the messy state of change; it exposes as well both the fault lines that made those cracks and the degree to which such paving could never be permanent.[2]

Dekker nevertheless seeks at least narrative repair as he quickly introduces a potential champion, someone to mend the path and pave the way forward:

> The losse of a Queene, was paid with the double interest of a King and Queene. The Cedar of her government which stood alone and bare no fruit, is changed now to an Olive, upon whose spreading branches grow both Kings and Queenes, Oh it were able to still a hundred paire of writing tables with notes, but to see the parts plaid in the compasse of one houre on the stage of this new-found world!
> (sig. C1v)

King James VI of Scotland brings to England a royal spouse and children, too, signs of stability that the former queen had not produced. The royal family is the promise of "spreading branches" that will reach into the future, beyond his reign, in generations of a continuous family, faith, and government. And in this way the king is the father of the nation he has only recently adopted, his "fruit" filling an England once "wildernes" that in so short space is instead interpreted as a "new-found world." Extending this point by way of biblical and anagogical terms, Dekker also represents the Stuarts as ushering in a new world that has followed a near-apocalyptic destruction of the old.

On the heels of the queen's passing and the promise of a new king, however, is England's most prolonged visitation of plague, which in Dekker's view is among the progeny of this "alteration of a State" that is an earthquake—reminding us perhaps of the earthquake and plague combination that alters Verona's communal and political terrain in *Romeo and Juliet* treated in the prior chapter. The plague by this way of thinking is one more Titan, one more rumbling reminder of what has been buried under tidy narratives. Dekker represents the plague by way of its polytheistically metaphoric and downright Ovidian potential, as well as by description of its material causes, treated at greater length in chapter 4. Here Dekker moves to prose to describe yet an ever more encompassing threat, which Charles Whitney rightly denotes as ecological in scope.[3] Dekker struggles to give words to the experience of radical reversals in circumstance:

> A stiffe and freezing horror sucks up the rivers of my blood: my haire stands an ende with the panting of my braines: mine eye balls are ready to start out, being beaten with the billowes of my teares: out of my weeping pen does the inck mournefully and more bitterly than gall drop on the palefac'd paper, even when I do but thinke how the bowels of my sicke Country have bene torne.
>
> (sig. C3r)

The plague extends the alteration of state, threatening to turn the nation from wilderness to still worse as a graveyard. Dekker carries a burden as witness that itself swells like a body that has become host to pressurized air "billow[ing]" until it forces passage out of its bodily prison. The symptoms of its physiological presence are those with which we are familiar from the preceding chapters of this volume: "haire stand[ing] an ende," "eye balls . . . ready to start out." What happens next is remarkable, as Dekker's passion spreads to his "weeping pen," such that its "inck" becomes literally a physiological symptom of his condition. The pen is an extension of himself, an appendage of body and mind, just as he, his pen, and the other human witnesses are extensions of their "sicke Country" with her "bowels" "torne" by plague hot within her. The event of the plague is an assemblage composed of these living and nonliving, human and non-human actants. In Dekker's conception, the experience is one of just such mixture, just such an "interinanimated" suffering that the country's suffering becomes Dekker's and then becomes the pen's, which spills on the page.[4] It is as if the country's suffering spreads to Dekker's readers, who then join the assemblage of the plague event, of the earthquake, of the ecological alteration of state.

Dekker next reveals a larger set of actants in the assemblage, as the delivery of the grief through his pen cannot be undertaken alone: "*Apollo* therefore and you bewitching silver-tongd Muses, get you gone, Invocate none of your names: Sorrow & Truth, sit you on each side of me, whilst I am delivered of this deadly burden." No traditional muse, nor apparently even the holy spirit such as Milton employs, will do for such a "deliver[y]" of Titanic, recalcitrant grief. Then Dekker calls again for additional aid:

> you the ghosts of those more (by many) then 40000. that with the virulent poison of infection have bene driven out of your earthly dwellings: you desolate hand-wringing widowes, that beate your bosomes over your departing husbands: you wofully distracted mothers that with disheveld haire falne into swounds, whilst you lye kissing the insensible cold lips of your breathlesse Infants: you out-cast and downe-troden Orphanes, that shall many a yeare hence remember more freshly to mourne, when your mourning garments shall looke olde and be for gotten; And you the *Genij* of all those emptyed families, whose habitations are now among the *Antipodes:* Joyne all your hands together, and with your bodies cast a ring about me: let me behold your ghastly vizages, that my paper may receive their true pictures: *Eccho* forth your grones through the hollow truncke of my pen, and raine downe your gummy teares into mine Incke, that even marble bosomes may be shaken with terrour, and hearts of Adamant melt into compassion.
>
> (sig. C3r–v)

The length of the passage in total, all parts combined, conveys a building up of emotion that must but cannot easily be discharged. In the form of "grones" 40,000 strong, his burden will only painfully be delivered through the "hollow truncke" of his pen and drop by drop of "gummy teares" into his ink. Those groans and tears will become for a time the text of *The Wonderful Year*, sure to "melt" even the very hardest "hearts of Adamant," a substance about which we know much from chapter 3. The telling, even with aid, eventually so wearies Dekker that he changes course several times, moving finally to a bleak form of satire, ending the work *sans* hope. For a time along the way, he is able to perform a kind of vociferation within a vital community engaged together in trauma: author, pen, ink, paper, nation, Sorrow, Truth, the dead, witnesses, and readers. This is a viable prescription for plague-time comfort.

A year later, Dekker again put to potent use the representation of the alteration of a state by way of an earthquake. In *The Whole Royal and Magnificent Entertainment*, Dekker joined with William Harrison, Ben Jonson, and Thomas Middleton to celebrate James I.[5] In its new context of celebration, they employ an earthquake via simile rather than as metaphor, and the earthquake is interrupted:

> The sorrow and amazement that like an earthquake began to shake the distempered body of this island, by reason of our late sovereign's

> departure, being wisely and miraculously prevented, and the feared wounds of civil sword (as Alexander's fury was with music) being stopped from bursting forth by the sound of trumpets that proclaimed King James, all men's eyes were presently turned to the north, standing even stone still in their circles, like the points of so many geometrical needles, through a fixed and adamantine desire to behold this forty-five years' wonder now brought forth by time.
>
> (229)[6]

Here too, Elizabeth I's death leads to passions stifled by shock until they "like an earthquake began to shake the distempered body of this island" for want of passage. But the trembling is abruptly as well as "wisely" and "miraculously prevented" by the arrival of James VI of Scotland. What Dekker, Harrison, Jonson, and Middleton suggest as an end to this quake is "the sound of trumpets that proclaim[ ]" the king. It is as if the trumpeted proclamation of James is a counterblast, able to compete with and immediately silence the rumblings that otherwise would have led to a "bursting forth" of "civil" disruption. As noted in prior chapters, the sound of heralding trumpets was regularly in this period associated with the sound of thunder, with that preceding earthquakes, and with the roar of lions. Here the trumpets become James' Jove-like weapon against the Titans of rebellion and perhaps against plague. This theory of replacing one powerful force with another was commonplace in plague writing by way of Galenic medicine that held a body can only host one major disease at a time. The plague regularly pushed syphilis or the stone out of the body, just as it was thought a person can only detect one odor at a time so that the smelling of a pomander or creating smoke in a chafing dish become ways to push away the stench and thus the threat of plague.[7]

James' trumpet blast also fossilizes all bodies in England into a "fixed and adamantine desire." This is a call quite different than that Dekker issues in *The Wonderful Year* in which all adamant hearts should melt. Here, all hearts are to become adamantly hopeful. By way of an interesting juxtaposition of unruly and "fixed" passions, of mutability and stability, the writers show James himself to be the "wonder" that moves such desire. He is the meteorophysiological hero able to subdue any threat. In their hands, James is more specifically the "forty-five years' wonder brought forth by time," as if the child of Elizabeth I's 45 years of reign—as if he had always been her heir, waiting, no less than within her or at least within the nation's imagination, to be born.[8] He was, after all, only 37 years old when he became King of England. His age in light of Elizabeth's years on the throne together aptly support of Dekker's description. The extended simile reinforces this representation of James' timely advancement:

> At length Expectation (who is ever waking and that so long was great) grew near the time of delivery, Rumour coming all in a sweat to play

> the midwife, whose first comfortable words were that this treasure of a kingdom, a man ruler, hid so many years from us, was now brought to light and at hand.
>
> (3)

Like the groans birthed through Dekker's pen in plague-time, James is born of "Expectation" that had been germinating if not pent up in the kingdom. An unusually beneficial "Rumour," with her "comfortable words," brings James forth as a meteor, but he is not as Orgoglio or as Queen Margaret, full of hot air; he is the "treasure," like Thaisa delivered from the sea and like diamonds long incubated in preparation for fully formed birth in an already perfectly mixed state. James was likewise born of pressure and tribulation—in the wake of the death of Elizabeth and during the plague visitation that the authors of the royal entertainment choose, wisely under the circumstances, to elide.

The year of 1603 was one of extremes for England's newly anointed king. In some estimations, including Dekker's, it was nothing less than "epochal"—a term I borrow from Van Kelly in his examination of epic "heroes [who] witness major epochal changes, from the Hellenic to the Roman world" for example.[9] The people of England at the turn of the century saw their own stories bearing this kind of resemblance to those of the Trojans, Greeks, Romans, and Geats. The England they knew under Elizabeth I had ended. At the dawn of the seventeenth century, with this new king was much that rewarded the hope of those who had held it; there was also much that reinforced fearful anticipation, making for a proverbial best and worst of times. A new king was on the throne, but first, and then for years thereafter—up to a full half of the kings' reign—plague ruled in his place.[10] Moreover, as Mary Thomas Crane, Kristen Poole, and Katherine Eggert have most recently documented, the emerging shift from a humanist, cosmological paradigm to a new Copernican, sun-centered system contributed to this sense that one was approaching the edge of such epochal change.[11] Writers like Dekker who captured the concerns attendant up on national change found themselves in many ways living through its initial eruptions, with what would be sure and more severe aftershocks to come. Unlike the Beowulf poet, Homer, or Virgil, however, these writers personally experienced the subjects about which they wrote—and some more directly than others, as poet and Dean of St. Paul's John Donne did.[12]

*

It is perhaps little wonder that from within such a context, John Donne imagined and experienced life in meteorophysiological terms. Donne, more than others, expressed this condition in deeply personal rather than fictionally representational ways. It appears that in his early years as a secular poet, he perceived himself as the meteoric child of polytheistic earth, one of her many metamorphic and passionately engendered progeny. Later in his life, as Dean of St. Paul's and one who had fathomed the nature of the human condition in which "Any Man's *death* diminishes *me*, because I am involved in *Mankinde*,"[13] Donne perceived himself as more stable in construction, with a body that would join with his soul for eternity in still more stable form and realm. Yet, even then, there are signs that he was troubled by the *meteora* his own body could send forth, as if he were earth, troubled by his progeny and unable to dissociate his soundness from their Titanically eruptive behavior. In *Devotions Upon Emergent Occasions*, he records these bodily disruptions as "these *earthquakes* in him selfe, sodaine shakings; these *lightnings*, sodaine flashes; these *thunders*, sodaine noises; these *Eclypses*, sodain offuscations, & darknings of his senses; these *blazing stars* sodaine fiery exhalations; these *rivers of blood*, sodaine red waters."[14] Even this man of faith finds himself distressingly unsure of himself—including of his literal, physical solidity—at such times of change.

Throughout his career, Donne was drawn to the contemplation and textual representation of all bodies trembling toward rupture and rapture. Such conditions were early for him provocatively thrilling. They supplied him with content and a volatile expressiveness to fuel his intense poetry. As I have suggested, with age came concern about these changing states—body, mind, soul—that were once so enthralling. As I demonstrate in the balance of the chapter, consideration of Donne's representations of meteorophysiological changeability offers a new vantage point from which to understand both Donne's literary expression over the course of his career and from which to continue our exploration of the early modern meteorophysiological condition of the trembling body.[15]

Donne's early writing shows him claiming changeability as at least a somewhat desirable condition and certainly as a marker for himself distinguishing him from others. In "The Calme," one of his very earliest poems dating to 1597 when he sailed with a fleet under direction of Queen Elizabeth I's favorite, Robert Devereux, Earl of Essex, Donne captures the sense of being trapped aboard a ship that has lost its wind-driven propulsion:

> Our storme is past, and that storms tyrannous rage,
> A stupid calme, but nothing it, doth swage.
> The fable is inverted, and farre more
> A blocke afflicts, now, then a storke before.
> Stormes chafe, and soone weare out themselves, or us;
> In calmes, Heaven laughs to see us languish thus.
> As steady'as I can wish, that my thoughts were,
> Smooth as thy mistresse glasse, or what shines there,
> The sea is now. And, as the Iles which wee
> Seeke, when wee can move, our ships rooted bee[16]

Donne's "stupid calm" is ironically worse than a "storm's tyrranous rage," and for a hyperbolic moment, the speaker believes that "Heaven laughs to see us languish thus." It is the "steady" course that is oppressive, causing the "languinsh[ing]." This condition of stasis, moreover, is made worse by the speaker's less than "steady" thoughts, which presumably run a faster pace than his physical body.

Donne then diagnoses the condition of calm in overtly distinctly meteorological terms, borrowing from materialist earthquake theory:

> Earths hollownesses, which the worlds lungs are,
> Have no more winde then the upper valt of aire.
> We can nor lost friends, nor sought foes recover,
> But meteorlike, save that wee move not, hover.
> Onely the Calenture together drawes
> Deare friends, which meet dead in great fishes jawes.

When "Earths hollownesses, which the worlds lungs are,/Have no more winde then the upper valt of aire," this runs counter to the regular circulation of air about and through earth. It is a condition for a wonder that signals other ill effects, such as the inability of the sailors to "recover" either "lost friends" or "sought foes." Ships and sailors, non-humans and humans are "meteorlike, save that wee move not." They all can only "hover" unnaturally, like an unnatural comet or earthquake stalled in its action, poised for eruption but then frozen.

Donne expresses exasperation over this appalling, unnatural condition for military men. It is a situation that Marlowe would have his Tamburlaine and Shakespeare would have his Glendower, Hotspur, and Henry V recognize as militarily weak. Donne explains the galling nature of his experience this way:

> Fate grudges us all, and doth subtly lay
> A scourge, 'gainst which wee all forget to pray,
> He that at sea prayes for more winde, as well
> Under the poles may begge cold, heat in hell.
> What are wee then? How little more alas
> Is man now, then before he was? he was
> Nothing; for us, wee are for nothing fit;
> Chance, or our selves still disproportion it.
> Wee have no power, no will, no sense; I lye,
> I should not then thus feele this miserie.
>
> (60–61)

The "scourge, 'gainst which wee all forget to pray" is the "Calme," that should bode well but here makes men "Nothing" and, for emphasis, "nothing fit" with "no power, no will, no sense." For this speaker, this is death. He would prefer to be the meteor that is the earthquake, the comet, the tempest—the soldier who finds himself meteorophysiologically animated for war, actively transformed in Marlowe's words by "windy exhalations,/Fighting for passage."[17] Donne is also describes a condition the more maddening for its being irremediable and, more specifically, for its being a circumstance that no amount of technology can amend. None of the best efforts at cartography or navigation can help one avoid or then remove oneself from such a condition.[18] This is a reminder that all human action is dependent on the elements in motion.

In "Elegy 8, The Comparison" (1594/95), Donne depicts the opposite of the calm, representing his far more desired meteorophysiological condition of changeability.[19] Describing his lover, Donne's speaker compares her mouth to another meteorological phenomenon:

> Thine's like the dread mouth of a fired gunne,
> Or like hot liquid metalls newly runne
> Into clay moulds, or like to that Aetna
> Where round about the grasse is burnt away.
> Are not your kisses then as filthy, and more,
> As a worme sucking an invenom'd sore?

The lover is earth that produces "Aetna"-like sulfuric flows in "hot liquid metals" from her mouth, which is both the "dread mouth of a fired gunne" and the proverbial hellmouth all at once. She is also a "worme sucking" as much as spewing poison that flows both ways. This is a body in motion, enticing because so, despite the obvious risk involved in approaching it.

Metonymically by her hand, Donne gives words to the speaker who elaborates upon his subject's enticing mutability:

> Doth not thy fearefull hand in feeling quake,
> As one which gath'ring flowers, still feares a snake?
> Is not your last act harsh, and violent,
> As where a Plough a stony ground doth rent?
> So kisse good Turtles, so devoutly nice
> Are Priests in handling reverent sacrifice,
> And nice in searching wounds the Surgeon is
> As wee, when wee embrace, or touch, or kisse.
> Leave her, and I will leave comparing thus,
> She, and comparisons are odious.
>
> (150)

She is "fearefull"—in the sense both to be feared and, herself, afraid to proceed in the erotic encounter, "As one gath'ring flowers, still feares a snake." The point is not to be "devoutly nice" but to "search[ ] wounds," share the kind of vulnerability that is dangerously, materially, and possibly permanently transformative. The "burn[ing]," "ren[ding]," poisonous activities come with this kind of "embrace, or touch, or kiss." It is the more erotically desirable thus to "quake" on the edge of metamorphosis, to be uncertain about what the probing of wounds will illuminate. The speaker desires exchange with another who can provide what Donne later, famously, calls "interinanimation"—change constant, mutual, and always new. It is as if the pair of lovers is together an assemblage that is meteoric in its constantly intermingling variability and reminders of contamination and death. It seems that the young Donne was unafraid to dare the biblical proverb, to hold fire in his bosom—an effort we know from chapter 4 that is meteorophysiologically dangerous to sustain.[20]

Five years later, Donne expresses his meteorophysiological changeability in different but equally poignant terms, rendering it again disconcertingly desirable as a condition. Writing "To the Countess of Bedford at New Year's Tide" (1610?), Donne describes himself this way:

> This twilight of two yeares, not past nor next,
> Some embleme is of mee, or I of this,
> Who Meteor-like, of stuffe and forme perplext,
> Whose *what*, and *where*, in disputation is,
> If I should call mee *any thing*, should misse.[21]

It is from this place of constant, uncomfortable, random change, "of stuffe and form perplext" that Donne hovers, worries, wonders, and writes. He cannot, he says, name even his own essential contents. Is he more air or earth, more man or woman, more meteor or human: "call mee *any thing*, should miss." Casting himself as "meteor-like," he is a changeable event rather than "*any thing*," and perhaps he was therefore the more attractive as a poet, more attractive to patronnesses who sought him out as a writer because of this very feature of his writing, this very changeability of persona and also of form, as each of Donne's poems offers twists and turns and surprises. "Protean innovator" is the term John Cary uses to describe Donne. Carey explains that he had an "imaginative commitment to instability," and in many ways, this served him well.[22] Yet in as many ways, this instability, this being in-between for Donne was not comfortable enough for him. It was not, for example, a positive version of the in-between human celebrated by Neoplatonism. Ficino posits humans as able to be angels or beasts along a continuum and as determined by will. Donne's energetic, exasperated poetry denies such simple power to the will, as later would his expressed Protestantism as Dean of St. Paul's. Later in his life, in this capacity, he was unable to embrace fully either his changeability or faith in post-resurrection unity of body and soul. He was too much a reader of the physical universe to settle on either side likely too attentive to loss—of his wife and children, of his own physical functioning, of the losses of his parishioners. He was surely too embodied ever to think comfortably beyond the body when it came to the fashioning of his identity.[23] And in this body, he observed constant psychophysiological change resembling that touching all other sublunary bodies, human and non-human.

*

Donne's most developed and intentionally troubled treatment of the meteoric human body appears late in his career, in his *Devotions Upon Emergent Occasions* (1624). It is also in this series of prose meditations that his most notable lines on human connection appear.[24] It is here that an earthquake becomes "the alteration of a State" in the dire terms that Dekker expresses. As a set, the *Devotions* represent a series of progressions and regressions of faith over days of illness. In movement from each of 24 Meditations to each Meditation's corresponding Expostulations and then Prayers is also both a movement from fear to contemplation to faith, which is common to the genre of Christian meditation. As a whole, the *Devotions* also show a movement through the interpretation of meteorophysiology from polytheistic and materialist to Christian foundations, building one upon the other and leaving none behind. *Meteora*, illnesses, and all sudden, radical, material alterations finally become for Donne signs of the fall of the sublunary system when Adam and Eve disobeyed God; they also become signs of assurance in salvation.

This comfort from the full view of *Devotions*, however, seems never able to endure, disintegrating each time it is rebuilt. In fact, he opens *Devotions* with fear for life and for soul. We hear Donne's compelling expression of this fear immediately at the start of Meditation 1, as he watches his body change, one alarming symptom of ill health after another: "VAriable, and therfore miserable condition of Man; this minute I was well, and am ill, this minute. I am surpriz'd with a sodaine change, & alteration to worse, and can impute it to no cause, nor call it by any name" (1–2). The opening word "VAriable" stands almost as a sentence unto itself, and the phrase of which it is a part sets the tone for all devotional meditations that follow. In many ways, this line acts as their brief abstract: "VAriable, and therfore miserable condition of Man." Not because sinful but because "VAriable" are humans "miserable." This is quite an assertion, in contrast with those made directly and indirectly by many of the writers treated in prior chapters of this book.

Donne's repetition of "this minute" delineates the special horror of this variability, as does the sentence housing the claim. By its construction, one can read "well" and "ill" as occurring in the same "this minute." This change is also, Donne says for emphasis, "sodaine" such that "I am surpriz'd" by it. The change is all the worse still for its originating by "no cause" and for the impossibility of naming it, as it is in each moment changing again like those imperfectly mixed *meteora* of Fulke's categorization. The implication is that these changes might be merely accidents, on the order Michael Witmore explains.[25] Donne is aware of this problem as he writes from what could be his deathbed of the possibility that these changes he witnesses and feels and against which he cries are ultimately meaningless beyond their temporary, mundane reality.

Still worse is that one can labor to prevent such change and still suddenly find oneself constituted by it, as he explains:

> We study *Health*, and we deliberate upon our *meats*, and *drink*, and *Ayre*, and *exercises*, and we hew, and wee polish every stone, that goes to that building; and so our *Health* is a long & a regular work; But in a minute a Cannon batters all, overthrowes all, demolishes all; a *Sicknes* unprevented for all our diligence, unsuspected for all our curiositie; nay, undeserved, if we consider only *disorder*, summons us, seizes us, possesses us, destroyes us in an instant.
>
> (2–3)

This is a passage worth reading aloud as it shows us Donne, the mesmerizing deliverer of sermons, able in a few lines to impart to his readers—with intentional use of that word "impart" as examined in chapter 3—the urgency of the situation to his readers. Emphasizing the material threat of the emergent occasion, Donne represents the human body as "that building" of health that is "a long & a regular work," fashioned stone by stone, one carefully-placed Galenic non-natural (diet, exercise, etc.) after another. Donne's body is not Queen Margaret's from chapter 4; it is one by which its agent has diligently acted for the securing of health. Even that carefully tended body is subject to sudden, utter "demoli[tion]." Donne's polysydetonic syntax underscores the urgency of the situation, and his antistrophic "all" gives resonant voice to the physical destruction as the "Cannon batters all, overthrowes all, demolishes all" that one had labored to secure. There is, he implies, little reason to trust in the constancy of one's bodily form. It can be returned in an instant to the elements of which it is most basically constituted, and these elements can be reconfigured, re-assembled into new forms that may not satisfactorily resemble one's body in the afterlife and may instead help to compose other sublunary material things, perhaps Hamlet's "politic worms," or a meteor (4.3.20), both of which would seem—the more disturbing yet—as spontaneous in their generation.

And yet, as much as Donne fears their meaninglessness, these changes make the body and mind of the sufferer alert, particularly as these changes might be the last he or she experiences: "and our *dissolution* is conceived in these *first changes*, *quickned* in the *sicknes* it selfe, and *borne* in *death*, which beares date from these first changes" (5). It is impossible not to imagine the outcome of these symptoms. Unfortunately, Donne does not find in them what Nausea via Fleming does in chapter 2—reason to anticipate the Apocalypse. For all of his desire to generate meaning from these changes, Donne finds in them only the promise of horrifyingly swift destruction to the point of "*dissolution.*"

Making matters worse, if that is possible, the progression toward death is not quite as hasty this or as the earlier "in a minute" suggests. The physical suffering has pronounced psychological accompaniments: "we . . . are pre-afflicted, super-afflicted with these jelousies and suspitions, and apprehensions of *Sicknes*, before we can cal it a sicknes" (4). The mutable human constitution can be largely balanced this minute but battered the next, or even in the same minute, and because we anticipate this change, we are afflicted already in advance, in the minutes prior to "this minute." Each possibly irregular pulse or other seemingly atypical bodily emission might be the sign of the physiological earthquake that will irremediably alter the terrain. Here as elsewhere, Donne gives voice to aspects of the vital materialism that Jane Bennett has recently articulated, wherein the human being is himself or herself an "assemblage," as "each member-actant maintains an energetic pulse slightly 'off' from that of the assemblage, an assemblage is never a solid block but an open-ended collective, a 'non-totalizable sum' with 'a finite life span'" (24). This, however, is Donne's fear—that inconstancy is the human condition. This dominant trait fails to square with the promise of the resurrection body and soul, and it opens the door to doubt.

Postlapsarian humans are in all these ways for Donne more meteoric than angelic, more sublunary than superlunary in their constitutions and motions. This is a point Donne extends forcefully, inquiring by way of the next metaphor:

> Is this the honour which Man hath by being a *litle world*, That he hath these *earthquakes* in him selfe, sodaine shakings; these *lightnings*, sodaine flashes; these *thunders*, sodaine noises; these *Eclypses*, sodain offuscations, & darknings of his senses; these *blazing stars* sodaine fiery exhalations; these *rivers of blood*, sodaine red waters?
>
> (5–6)

Donne's body is a "litle world" in that both his body and earth undergo material alteration and produce "exhalations" and other material symptoms of change. But what, then, is the "honour" in such a condition? The *meteora* of the sublunary system are unstable and lacking in form, as soon dissolved as created; so too, Donne suggests, are the emergent occasions of the human body, these "shakings," the fainting, and profuse bleeding. These *meteora* of the body underscore the material, sublunary composition of humans who are subject to the natural forces governing all sublunary matter—forces that may have no explanation outside of their material and efficient causes.

By syntax Donne further builds the case that the misery of the human condition is its variability, as he repeats "sodaine" to give voice to each percussive cannon blast upon the body. "Is this the honour . . . [?!]" he asks, as if, given the meteoric performance of his eruptive body, the very thought of "honour" is galling. In many ways, Donne offers a direct response to Helkiah Crooke's positive proclamation on the resemblance between world and body in *Mikrokosmographia a description of the body of man* (1615). I treat his proclamation at some length for its thoroughgoing meteorophysiology and for its contrast with Donne's expression of his experience in *Devotions*. Imitating natural philosophers, Crooke explains in detail the basic materials and motions of the sublunary sphere, with the four elements and the four humors combining to create bodies that are either "simple" and "perfect" in form, and thus "living," or that are "mixt" and "imperfect"—the latter being "Meteors," about which he announces invitingly:

> Behold also, the wonderfull Analogie of the Meteors of this little world. The terrible Lightning and fiery flashes and impressions, are shewed in the ruddie suffusions of our eyes when we are in a heate and furie, as also by those or darting beames which we throw from the same. The rumbling of the guts, their croaking murmurs, their rapping escapes, and the hudled and redoubled belchings of the stomacke, do represent the fashion and manner of all kindes of thunders. The violent and gathering rage of blustering windes, tempestuous stormes and gustes, are not onely exhibited, but also foreshewed by exhaled crudities, and by the hissing, singing, and ringing noises of the eares. The humor and moistnesse that fals like a Current or streame into the empty spaces of the throate, the throtle and the chest, resembleth raine and showers. Thicke and concocted Flegme, that comes up round and roundly when we Cough, carries the likenesse of Haile-stones; teares do represent the Dew: shaking, shrinking, trembling, & throbbing motions, resemble the Earth-quakes. There are also found in our bodies, Mines and quarries, out of which, Mettals and stones are digged, not to builde, but to pull downe the house; so the stones of the Kidneyes and bladder do carry a resemblance of Mines and Mineralles.[26]

This "Analogi[c]" relationship exposes so exact a match, the microcosm to the macrocosm, that Crooke recites these meteorophysiological symptoms in order of the accepted materialist categories for *meteora*, "Fiery, Aiery, Watry, Earthy" (7), starting with the highest, lightest, and hottest, working down through the sublunary sphere to the lowest, heaviest, coldest with earthquakes and minerals. Every symptom of the body has, for

Crooke, its natural meteorological equivalent, and we recognize many of them resonantly even if not exactly matched from within texts treated in prior chapters of this volume. Underscoring the relationship between these bodies, the margin note reads "The meteorologie of the little world" (7).

It is as interesting that Crooke rather drifts from some of the alignments between physiology and meteorology that Aristotle, Pliny, Seneca, Fulke, Fleming, and others supply. For example, although these writers mention directly or indirectly a relationship between thunder and earthquake, they would have associated with earthquakes the bodily symptoms that Crooke associates with thunder—the rumbling and so forth. Crooke's aligns his gentler throbbings with the quake, thus rendering quakes so benign as to suggest he does not view their potency as others, including Donne, certainly did. More than 30 years after the 1580 earthquake, it is possible that this Englishman had never felt his body shaken by an earthquake thus to distinguish it from the lesser rumbling of more frequent thunder.

Explaining in very general terms, then, the macrocosmic-microcosmic relationship between bodily changes, human and world, Crooke returns to the concept of mixture. Resemblance to Donne's representation of earthquakes and of meteorology further pales:

> This is the Meteorology of this Little worlde, this is the demonstration of those things therein that are imperfectly mixed. And if you require in man an example of a bodye perfectly mixed, behold and consider the whole body; in which, there is that concord and agreement of the foure disagreeing qualities, and so just & equal a mixture of the elements, as that it is the very middle and meane amongst all living and animated things. This Little World therefore, which we call Man, is a great miracle, and his frame and composition is more to be admired and wondered at, then the workemanship of the whole Universe. For, it is a farre easier thing to depaint out many things in a large and spacious Table, such as is the world; then to comprehend all things in one so little and narrow, as is the compasse of mans body.
>
> (8)

Crooke's purpose is pronounced: he treats meteorology primarily to describe the condition of the human body and to celebrate the superiority of the human over all things in the "whole Universe." Unlike the sublunary world itself, Crooke finds the human body "perfectly mixed" in composition of elements. This human body alone is able, he asserts, to manage its many *meteora*, containing and balancing them by its being "the very middle and meane amongs all living and animated things" such that its "frame and composition" is the more remarkable, because able to contain and balance all elements and motions perfectly. The human—in contrast to Donne's experience and claims—is "a great miracle."

Crooke had already shared his rationale for the entire work (at book 1, chapter 2); in it, we hear what may be some further cause for his slighting, or perhaps we might call it the taming, of earthquakes and other Titanic meteorophysiological shocks:

> AS the soule of man is of all sublunary formes the most noble, so his Body, the house of the soule, doth so farre excell . . . the measure and rule of all other bodies. There be many things which set foorth the excellency of it, but these especially among others. The frame and composition which is upright and mounting toward heaven, the moderate temper, the equal and just proportion of the parts; and lastly, their wonderfull consent & mutuall concord as long as they are in subjection to the Law & rule of Nature: for so long in them we may behold the lively Image of all this whole Universe, which wee see with our eyes (as it were) shadowed in a Glasse, or desciphered in a Table. (4)

In the four categories of "frame", "temper", "proportion", and "concord", By this time in his *Mikrokosmographia*, he says, the human body mirrors and exceeds the universe itself; that is, it does so "as long as they are in subjection to the Law & rule of Nature." Crooke exposes the rule to which very few bodies can adhere. Those bodies excessive in their heat, for example, cannot by this stipulation count as paragons of creation. We might think of the bodies represented by writers and treated in former chapters, those which are surely not by Crooke's standards "great miracle[s]" of "wonderfull consent & mutuall concord." These prior bodies register each of their meteorophysiological symptoms uniquely, as Donne does; they in no way experience these symptoms as uniformly held in comfortable relation.

To find all parts of the body always in concert rather than at least sometimes in conflict, is indeed a remarkable thing, because from what Crooke explains, such a condition depends on each symptom having essentially been flattened out to equalize them. Christian writers for centuries had done just this with sins and all sublunary disruptions, turning them all alike into signs of the Apocalypse. Similarly, though on the other end of a much larger spectrum of human distinction from Christian to materialist paradigms, the fear, allure, and potential for all sublunary things to be equal and interchangeable by extreme materialist theory bordering on atomism which threatened in the way a plague pit did, with nameless, faceless numbers of corpses like atoms, meaningless. Neither Donne nor others treated here would have been in a position o embrace in full the concepts of "human indistinction" or of "flat ontology," however much their fears were based on layered evidence pointing in these directions.[27] Like the characters of Queen Margaret and Britomart, Donne and others did experience and write of unique, surprisingly unpredictable symptoms of material change rather focus on the concord in their bodies and in the systems in which these bodies were embedded.[28] In Randall Martin's words, "Donne and contemporaries like Shakespeare were discovering what modern research has confirmed to be the planet's normal conditions: a world continually subject to geophysical ruptures, mutable climate patters, and species change."[29] Within the change nonetheless, and to underscore the point, writers in this period were desperate to find humans invested with some sort of distinction, even if only through grace or will. That does not mean they always could or did, and Donne expresses more directly than others the outright horror of having intentionally labored to advance himself, his building of health, one stone and one salubrious regimen after another, only to find his body's symptoms showing he is anything but "perfectly mixed" in Crooke's sense. Crooke's work, though interesting and in many ways learned, could only be for Donne impractical as a guide to meteorology and physiology.

True to his own relentless thinking process, which at the same time always admitted variability, Donne inquires further in *Devotions*, exploring the pain of radical change for which there is no apparent cause:

> Is he a *world* to himselfe onely therefore, that he hath inough in himself, not only to destroy, and execute himselfe, but to presage that execution upon himselfe; to assist the sicknes, to antidate the sicknes, to make the sicknes the more irremediable, by sad apprehensions, and as if hee would make a fire the more vehement, by sprinkling water upon the coales, so to wrap a hote fever in cold Melancholy, least the fever alone shold not destroy fast enough, without this contribution, nor perfit the work (which is *destruction*) except we joynd an artificiall sicknes, of our owne *melancholy*, to our natural, our unnaturall fever.
>
> (6–8)

Donne's thought experiment is extreme: perhaps one is a little world to bring about one's own "*destruction.*" And, then, worse than this, one can "presage" that destruction, too. The physical *meteora* of the body—including rapid pulse and excessive exhalation—are anticipatory signs of future trouble; thus they "assist" in co-conspirator fashion, in the causing of suffering. Given the belief at the time that fear and other negative passions, such as the "apprehension[ ]" of suffering, were able to weaken the body, opening it to illnesses when otherwise the body might better have maintained its composure, Donne's concern is medically well-founded.

Precisely versed in the humoral theory behind this belief, which we saw expressed in similar but more prescriptive detail by Thomas Elyot in chapter 4, Donne wishes to avoid such "sad apprehensions," the "artificial sicknes of . . . *melancholy*" that will "unnaturall[y]" augment the "natural" fever. And yet, as he understands it, it is as if "the fever alone should not destroy fast enough, without this contribution," as if illnesses can only reach their deadliest of stages with the help of the imagination. Donne cries out anaphorically again, this time in "Os" against his condition, the little world that no longer trusts its form: "O perplex'd discomposition, O ridling distemper, O miserable condition of Man" (8). This is the alteration of state—the very state of discomposure and fearful destructive trembling—that threatens people of faith who come to imagine themselves as *meteora*. This is the alteration of state that can similarly threaten those who know themselves to be assemblages and events more than as divinely created humans. It is the more terrifying, I would argue, to see oneself from the former perspective, no longer stable but always devolving just as one is becoming, and worse, never having had an intended evolution, but having been from the start imperfectly mixed of earth and water to which one will return. We are notably far from the kind of wonder that inspires Crooke when he contemplates his well-composed little world.

In Expostulation 1, Donne renders himself nearly in the condition of this very dissolution he anticipates. In keeping with his leading practice of expression in poetry and prose, however, it is at this lowest of places—as dust, as dissipated, dis-elemented meteor—that Donne turns abruptly to a full pronouncement of faith by way of another interrogative:

> IF I were but meere *dust* & *ashes*, I might speak unto the *Lord*, for the *Lordes* hand made me of this *dust*, and the *Lords* hand shall recollect these *ashes*; the *Lords* hand was the wheele, upon which this vessell of clay was framed, and the *Lordes* hand is the *Urne*, in which these *ashes* shall be preserv'd. I am the *dust*, & the *ashes* of the *Temple* of the *H. Ghost;* and what Marble is so precious? But I am more then *dust* & *ashes*; I am my best part, I am my *soule*. And being so, the *breath* of *God*, I may breath[e] back these pious *expostulations* to my *God*.
>
> (8–9)

Within four sentences, Donne moves from imagining himself "*dust* & *ashes*," the condition of earth at its most elemented and infertile, to "being so, the *breath* of *God*." This path traversed from "*dust*" to "*breath*," from earth to air, from death to resurrection, is short. It would be wrong to think it has been made without effort—and that being God's. The repetition of "the *Lordes* hand" reinforces the intent and labor that God employs to fashion humans from dust. This image of God molding a human clay recalls the human body that is the "building" of Meditation 1—the building that, Donne reveals, is, and perhaps has always been, the very "*Temple* of the *H. Ghost*." The body now, as reclaimed by way of what Donne implies is a memory, is made of "Marble, wccording to materialist meteorology via Fulke is among meteor of perfect mixture, with increased durability and utility.[30] Just so is the breath redeemed. Donne's "expostulations" become a part of the heavenly respiration between God and his creatures. They are a "breath[ing] back" to God, even as they are at once a kind of protest; they have meaning, value, and even a sustained existence beyond their expulsion from the body. They are as inspiring blessings to the exhausting curses of this volume's chapter 4.[31]

It would seem that Donne might now breathe easily, so to speak, saved as he appears to be from despair by having reconfigured his "apprehensions" into inspirations; his "expostulations" into respiration; and his *meteora* into signs of faith. But that would be unlike Donne, who instead extends his baroque meditation by another fold to challenge his newfound security. Why, he wonders, is such a sensitive system of apprehension built into the body but not built into the soul? In a question that would sound foolish in another context, he asks, "My God, my God, why is not my soule, as sensible as my body?" (9–10). And he means this rather literally, extending the query with details: "Why hath not my *soule* these apprehensions, these presages, these changes, those antidates, those jealousies, those suspitions of a *sinne* as well as my body of a *sicknes*?" (10). Why hath not the soul earthquakes, roars, heralding trumpets? If the soul had those *meteora*—the eruptive warnings of impending destruction—it might be better able to avoid sin. Instead, lacking such a monitoring system, one charges headlong to meet spiritual threats: "I breake into houses, wher the plague is" (11), he says. Having wished his body lacked such torturous apprehensions, now he yearns for his soul to have them too, the better to protect himself from himself, from himself as "a reciprocal plague"—a self-title he employs in a sermon from the same year.[32]

Approaching another edge of exasperation and what looks to be one of the many spiritual relapses he begins to experience in concert with physical ones as the *Devotions* continues, Donne checks himself and instead reinforces his faith. He does so by turning to the Bible, remembering Job and Jacob, biblical archetypes of the suffering servant who were similarly challenged by emergent occasions that threatened physical and spiritual undoing. God may not have "imprinted" in us our "miserable condition," by which he means the original sin that causes the eventual dissolution of the body, but he *has* "imprinted a *pulse* in our *Soule*." This is the desired warning system; it was always here. One must only learn to "hearken unto it," in part by refusing to "talk," "jest," "drinke," or "sleepe it out" (13). Essentially, listening to or feeling for this pulse—for the pulse that is steady rather than for each of the "thousand natural shocks"—requires such attentiveness that the most basic of human activities, such as the acquisition of the essential Galenic non-naturals of food and sleep, and perhaps the taking of one's *physical* pulse, can distract one from perceiving the soul's informing rhythm.

Once one recognizes and makes the effort to attend to that spiritual pulse, however, detection can be certain. As Donne explains, the perception of the spiritual pulse shows it to be not only steady but also enduring, as rent paid when God is the tenant: "not yearely, nor quarterly, but hourely, and quarterly." With a return to the fearful "minute" of impending dissolution but with a difference, with the pulse of the soul, "Every minute he renewes his mercy" (15). This "minute" of certainty can entirely overwrite the "in a minute" and "this moment" associated with the physical threats from the preceding Meditation. Here "*[e]very minute*" is and will be the invitation to be "*heale*[d]." And each of these minutes, as each pulse of the soul, gives promise of an informing principal in humans that has already imprinted and prepared them for salvation, body and soul.

*

In the concluding Prayer of this first Devotion, the "*earthquake*[ ]" of the Meditation that becomes the predictable "pulse" of mercy in the Expostulation takes what Donne will show to be its original form as the very voice of God: "Thy voice received, in the beginning of a sicknesse, of a sinne, is true health" (18). God's voice is the earthquake of the body, of the mind, and, later, of the bell that tolls, a terrible and wonderful sign coming "at the beginning," "at the approach," of destruction and of resurrection. It is the first symptom of change both of material dissolution but of God's salvific plan. Donne closes this Devotion by asking for the wisdom to interpret all symptoms and apprehensions of the physical body and of the soul as this voice of God, so that "vain imaginings" become "constant assurance" and every "falling" becomes a "fly[ing]" or rais[ing]." With such wisdom, with every minute comes the assurance of what Donne later, at the close of Expostulation 2, will explain this way: "Though I be dead I shall heare the voice the sounding of the voice, and the working of the voice shall be all one; and all shall rise there in a lesse *Minute*, then any one dies here" (34–35). Donne reworks the minute, moving it from threat to promise, from its standing as the last moment of embodiment before utter and final material dissolution to its sounding regularly through life as God's trumpeting "voice" of mercy that is also the sound of post-apocalyptic reconstitution. However many years, decades, or centuries pass between death and the body's resurrection, the individual will have the experience as if he or she were raptured immediately ("in a lesse *Minute*") after death. The experience after death will be as if one has bypassed the decay and dispersal of the body's portion of the process.

This is "the honour which Man hath by being a *litle world*, That he hath these *earthquakes* in him selfe, sodaine shakings." "[T]hese earthquakes," the radical *meteora* of the passionate body, constitute an eschatological pulse that must for humans be experienced uniquely one beat at a time. Humans, Donne implies, were not so constituted as Crooke would suggest, to be calmly enduring life in the hope of achieving miraculous balance or stoic *ataraxia*. Each quake and tempest is a distinct, literal event that the passionate body must experience, newly with each encounter, with each event's coming into being, persistence, and going. At the same time, for Donne, these pulses, these quakes, are each the unique wounds of Christ, authentically suffered in his human flesh.[33] Donne's "emergent occasions"—expressed not only in the first devotion but repeatedly throughout the set—are the fleeting, meteoric tribulations of the body and imagination, but they are also representations of the pulse, the breath, the voice, and the compassionate mercy of God.

As mentioned earlier in this chapter, Donne's devotional movement here from the Meditation to its Expostulation and Prayer is also both a movement from fear to faith, common to the genre of Christian meditation, and a movement through the interpretation of meteorophysiology from polytheistic and materialist to Christian foundations. *Meteora*, illnesses, and all sudden, radical, material alterations are signs of the fall of the sublunary system when Adam and Eve disobeyed God. At the end of the story, the post-resurrection faithful will become perfectly mixed and enduring *meteora* with bodies, minds, and souls stably conjoined in the experience of the immutable God with whom all creation then knowingly will resound and respire. Donne thus draws from Ovid, Aristotle, and the Bible to express the complex psychophysiological human condition that is always also meteoric in its materials and motions.

This is not to suggest that in *Devotions* Donne builds steadily or comfortably over time toward such assurance that, one day, sublunary, human *meteora* will be reconstituted as adamantine signs of Christian salvation. Donne is the first to share his own physical and spiritual relapses, which turn the golden treasure wrought from tribulation expressed in the meditation of the seventeenth Devotion back into reasons for the return of emergent occasions more virulent and thus fear- and doubt-producing than the first. It is with words on the gravest symptom of threatening relapse—"the most immediate exercise of that affection of fear"—that Donne closes his examination of his 23 days of illness (605). Donne's faith will be informed by unique, unpredictable experiences of fear and trembling, none of them preventable or easily named, all each time posing authentic challenge for each body, which will itself be uniquely assembled in each minute. By the end of *Devotions*, Donne *has* nevertheless shown a resilience of faith that can persist through and beyond each next emerging symptom and relapse, physical and spiritual. In this, he performs for his readers what I think might best be called a "practice of permanence"—even, and perhaps heroically, from within the human condition of fully embodied mutability, where, Donne tells us in Expostulation 3, "As yet God suspends mee betweene *Heaven* and *Earth*, as a *Meteor*" (56). Indeed, to return to the beginning of this volume, to the words of Shigehisa Kuriyama, Donne helps us experience this "[o]nce upon a time, [when] all reflection on what we call the body was inseparable from inquiry into places and directions, seasons and winds," when "human being was being embedded in a world."[34] For Donne to be embedded is still always an experience of impermanence and threatened change; it is to hover interactively, enmeshed in systems always in motion that by comparison renders Dekker's "alteration of a State" and the 1580 earthquake experience as child's play.

Notes

1. Thomas Dekker, *The Wonderful Year* (1603), sig. B2v.
2. Jane Bennett, *Vibrant Matter: A Political Ecology of Things* (Durham: Duke University Press, 2010), 119.
3. Charles Whitney, "Dekker's and Middleton's Plague Pamphlets as Environmental Literature," in Rebecca Totaro and Ernest B. Gilman, editors, *Representing the Plague in Early Modern England*, Routledge Studies in Renaissance Literature and Culture (New York: Routledge, 2011), 201–218.
4. As most will know, I draw the term "interinanimation" of course from Donne's use of "interinanimat[es]" in *The Ecstasy*, where there the assemblage of lovers also more clearly includes the souls of the lovers and not only their element-composed material bodies.
5. In Gary Taylor, John Lavagnino, and P. Jackson MacDonald eds., *Thomas Middleton: The Collected Works* (Oxford, UK: Clarendon Press, 2007).
6. Taylor, Lavignano, and MacDonald note being unable to find the meaning for this allusion to Alexander's fury and music (229), but Kate van Orden in *Music, Discipline, and Arms in Early Modern France* (Chicago: University of Chicago Press, 2005, 14) explains this is a reference to Alexander the Great's magnificently fine-tuned soul that could so quickly by music be re-harmonized out of battlefield action.
7. On the idea that it was only possible to have one major disease at a time, the more potent acting to block or subsume the weaker, see for example Holland, who says that the plague "may affright the rest [of the diseases] we nam'd before" (Rebecca Totaro, *The Plague Epic in Early Modern England: Heroic Measures, 1603–1721* [New York: Routledge, 2012], 163), and see Thomas Sprat's Pindaric ode, *The Plague of Athens* (1659): "Whatever lesser Maladies men had,/They all gave place and vanished;/Those petty Tyrants fled,/And at this mighty Conqueror shrunk their head" (10). For more on this text please also see Totaro, *The Plague Epic*.
8. I owe this observation on James' age to Gary Taylor.
9. Van Kelly, "Introduction: Criteria for the Epic: Borders, Diversity, and Expansion," in Steven M. Oberhelman, Van Kelly, and Richard J. Golsan, editors, *Epic and Epoch: Essays on the Interpretation and History of a Genre*, Studies in Comparative Literature 24 (Lubbock, TX: Texas Tech University Press, 1994), 20 n. 14. See also Rebecca Totaro, *The Plague Epic*.
10. On increased mortality in James' reign, see Paul Slack, *The Impact of Plague in Tudor and Stuart England* (London and Boston: Routledge & Kegan Paul, 1985; reprinted with corrections, Oxford: Clarendon Press, 1985; New York: Oxford University Press, 2000), 146–147—currently the preeminent source on British historical analysis of the plague; Charles Creighton, *History of Epidemics in Britain: From AD 664 to the Extinction of the Plague*, 2 Volumes (Cambridge: Cambridge University Press, 1891–1894), 1:229–303, 575; Charles Mullett, *The Bubonic Plague and England: An Essay in the History of Preventative Medicine* (Lexington: University of Kentucky, 1956), 31–210;

and *English Short Title Catalogue* listings under "England, Proclamations" for 1604–9, 1625, 1636.

11. See especially the introductory chapters to Mary Thomas Crane, *Losing Touch With Nature: Literature and the New Science in Sixteenth-Century England* (Baltimore: Johns Hopkins University Press, 2014); Kristen Poole, *Supernatural Environments in Shakespeare's England: Spaces of Demonism, Divinity, and Drama* (Cambridge: Cambridge University Press, 2011); Katherine Eggert, *Disknowledge: Literature, Alchemy, and the End of Humanism in Renaissance England* (Cambridge University Press, 2015); Mary Floyd-Wilson, *Occult Knowledge, Science, and Gender on the Shakespearean Stage* (Cambridge: Cambridge University Press, 2013).
12. On witnessing epochal change, see Kelly, "Introduction: Criteria for the Epic: Borders, Diversity, and Expansion," 20 n. 14.
13. John Donne, *Devotions upon emergent occasions and severall steps in my sicknes digested into I. Meditations upon our humane condition, 2. Expostulations, and debatements with God, 3. Prayers, upon the severall occasions, to Him / by John Donne . . . , London : Printed for Thomas Jones, 1624*, 416. See also John Donne, *Devotions XVII Meditation*, John Cary, editor (Oxford: Oxford University Press, 1990).
14. Donne, *devotions upon emergent occasions*, 6.
15. My use of "unwilled" reveals a debt here to Luke Taylor, "Donne's Unwilled Body," *John Donne Journal* 30 (2011): 99–121.
16. "The Calm," in *Poems, by J.D. With elegies on the authors death, London: Printed by M[iles] F[lesher] for John Marriot, and are to be sold at his shop in St Dunstans Church-yard in Fleet-street, 1633*. Folio pages 59–61. This poem is best paired with Donne's "The Storm," which deserves fuller attention in meteorophysiological context.
17. Christopher Marlowe, "*Tamburlaine the Great: Part One*," 1.2.47–51.
18. Special thanks are due to Gail Kern Paster for this keen observation.
19. *Poems, by J.D. With elegies on the authors death*, 149–150.
20. As concluded in chapter 4, when Proverbs 6:27 asks, "Can a man take fyre in his bosome, & his clothes not be burnt?" (Geneva), the resounding answer is "no."
21. *Poems, by J.D. With elegies on the authors death*, 87. See also among many other choices for meteorophysiological exam, see "A Valediction Forbidding Mourning," Holy Sonnet 9: "Oh, To Vex Me, Contraries Meet in One." An examination in this context of Donne's use of "elemented" in verse and prose is also needed. Among the few treatments is a short etymological study by Edgar S. Laird, "Love 'Elemented' in John Donne's 'Valediction: Forbidding Mourning'," *ANQ: A Quarterly Journal of Short Articles, Notes, And Reviews* 4.3 (1991): 120–122.
22. John Carey, "Introduction," in John Carey, editor, *John Donne: The Major Works*, Oxford World's Classics (2000; Oxford and New York: Oxford University Press, 2008), xxxi.
23. As Ramie Targoff explains, "If we were forced to identify Donne within a single philosophical school, it would almost certainly be Aristotelian" *John Donne, Body and Soul* (Chicago: University of Chicago Press, 2008).
24. So often quoted are these lines, and to powerfully do they register, that they have often been misquoted in poetic rather than prose form.
25. Michael Witmore, *Culture of Accidents: Unexpected Knowledges in Early Modern England* (Stanford, CA: Stanford University Press, 2001); Aristotle, *Meteorology*, H.D.P. Lee, translator (Cambridge, MA: Harvard University Press, 1952), 1:555.
26. Indebted to Susan Staub for this reference, I cite Helkiah Crooke, *Mikrokosmographia a description of the body of man. Together with the countroversies*

thereto belonging. Collected and translated out of all the best authors of anatomy, especially out of Gasper Bauhinus and Andreas Laurentius. By Helkiah Crooke Doctor of Physicke, physitian to His Majestie, and his Highnesse professor in anatomy and chyrurgerie. Published by the Kings Majesties especiall direction and warrant according to the first integrity, as it was originally written by the author (1615), 7–8. For more on Crooke's meteorophysiology, see Eve Keller, *Generating Bodies and Gendered Selves: The Rhetoric of Reproduction in Early Modern England* (Seattle: University of Washington Press, 2007), 52–56.

27. Jean E. Feerick and Vin Nardizzi eds., *The Indistinct Human in Renaissance Literature*, Early Modern Cultural Studies 1500–1700 (New York: Palgrave Macmillan, 2002), v. On the "flat ontology," see Manuel Delanda, *Intensive Science and Virtual Philosophy* (New York: Bloombury, 2013), 51. On the levelling nature of the early modern plague pit, with its "the pile of undifferentiated corpses," and its relationship to the pile of bodies on stage at the end of early modern revenge plays, see Michael Neill, *Issues of Death: Mortality and Identity in English Renaissance Tragedy* (Oxford: Clarendon Press, 1998), 9.
28. See on Donne and Francis Bacon, which deserves more treatment, Desiree Helleger, *Handmaid to Divinity: Natural Philosophy, Poetry, and Gender in Seventeenth-Century England* (Norman, OK: University of Oklahoma Press, 2000), especially chapter 1 in which she discusses the fact that "Donne problematizes any division between these two domains" of science and poetry (23). Please also see Mary Baine Campbell on Francis Bacon's treatment of earthquakes and volcanos, which obviously draws from the period's meteorophysiological understanding of the world more than from a scientific understanding of the sort we would label as such; making this point, she cites Bacon's Catalogue of Particular Histories by Titles, including this entry,

    > 18. History of the greater Motions and Perturbations in Earth and Sea; Earthquakes, Tremblings and Yawnings of the Earth, Islands newly appearing; Floating Islands; Breakings off of Land by entrance of the Sea, Encroachments and Inundations and contrariwise Recessions of the Sea; Eruptions of Fire from the Earth; Sudden Eruptions of Waters from the Earth; and the like.
    >
    > (Mary B. Campbell, *Wonder & Science: Imagining Worlds in Early Modern Europe* [Ithaca, NY: Cornell University Press, 1999], 79)

29. Randall Martin, *Shakespeare and Ecology* (Oxford: Oxford University Press, 2015), 5.
30. William Fulke, *A goodly gallerye* on the meteorological constitution of these materials: "The last sort namely earthly, *Meteores* are called perfectly mixed, because they wil not easely be chaunged and resolved from that forme which they are in, as be stones, metalles and other mineralles" [1563], sig. A1v).
31. The "expostulation" is itself a kind of purgation, a calling out to God in ways that might also suit the definition for what Thomas Elyot (*Castel of Helthe*) and others call vociferation—a calling out that is part of a physical regimen to rectify the body's respiration; see chapter 4. See also Shirilan on Burton's notion of "exoneration" in which one "disburden[s] himself by 'confessing his griefe to a friend'" (2:99), a kind of "'simple narration," in Burton's words cited by Shirilan, that "'many times easeth our distressed minde'" (150).

32. John Donne, "From an Undated Sermon (1624–25): A Reciprocal Plague," in John Carey, editor, *John Donne: The Major Works*, Oxford World's Classics (2000; Oxford and New York: Oxford University Press, 2008), 354.
33. On the importance of Christ's bodily suffering, and specifically as Man of Sorrows, see Michael C. Schoenfeldt, *Bodies and Selves in Early Modern England: Physiology and Inwardness in Spenser, Shakespeare, Herbert, and Milton*, Cambridge Studies in Renaissance Literature and Culture 34 (Cambridge: Cambridge University Press, 1999), 1–2.
34. Shigehisa Kuriyama, *The Expressiveness of the Body and the Divergence of Greek and Chinese Medicine* (New York: Zone Books, 1999), 162.

Afterword

Early modern meteorology was an unusual area of natural philosophy for many reasons, including the degree to which it was technologically incompatible with modern science, so little of the study viably conducted in a lab. The advance of meteorology requires technology that was far in the future, such as that related to the harnessing of electricity and that enabling sonar, radar, and seismography. At the same time, early modern meteorology was more readily compatible with modern science than other areas were, due to the direct and widespread necessities for and implications of the practice of meteorology across economies from households and professions to nations, especially as nations increased their international trade operations. In Craig Martin's words, "the necessary role of experience in developing meteorological theories led scholars to ancient texts where they found empirical evidence in chronicles and of past weather events" (2). They drew "from a number of fields" and this meant that "Renaissance meteorology was by no means static" (2) and was, by my findings, always blended with physiology. Renaissance meteorology—which, to drive home the point, is always meteorophysiological in representation—was dynamic and changing. Like its subjects, it provided a constantly renewing resource for wonder and for testing the limits of human potential and the imagination itself.

For this reason, there are many avenues for further examination of early modern meteorophysiology. Among the reservoirs of early modern writing that I hope to see investigated, and for which my initial findings assure reward, as well as outright surprise, are the following: writing by women across classes and through genres, particularly related to maternal experience; by gynecologists and other practitioners of natural philosophy on the ground, so to speak; by writers contemplating in complex ways a Lucretian meteorophysiology and working in concert with the three dominant paradigms, polytheistic, materialist, and Christian; by writers reporting on other radical meteorophysiological phenomena in the period and soon thereafter, such as meteo-tsunamis along the Channel and widespread regular flooding; by writers contemplating directly and indirectly spontaneous generation in its many forms material and immaterial; by those treating air and its chemical-military values as well as other related technological advances in the seventeenth century and beyond; by writers advancing utopian and science fiction genres; comparatively by medieval writers; and comparatively by those writing into the later seventeenth through nineteenth centuries.

Into the latter half of the seventeenth century, for example, early moderns continued to employ roughly the same conceptual systems to account for the *meteora*, such that a look into that immediate future of early modern meteorophysiology shows much that directly complements works already presented in this volume—the use of the same terminology and paradigms. With respect instead to realms of the imagination, to the parameters for the textual representation of human being, and to advances in technological intervention first posited imaginatively in literature, however, seventeenth century literature shows an expansion in the variations possible for and application of meteorophysiological experience. This is so across a range of writers, but I will focus briefly on those overtly treating the phenomena of world-making. In *The Blazing World* (1666), Margaret Cavendish depicts a quasi-utopian, fantastical world composed of fire and life forms previously unrepresented in English literature. Her earth is non-gendered and unresponsive, its elements controlled not by God but by the author, who, if she does not like the real world as she finds it, may create her own imaginary one, or two, or three. In *Paradise Lost* (1667), Milton aims to explain the ways of God to humans and thus account for sin and its punishment, but he also depicts a quasi-polytheistic mother earth, who is wounded and convulses physically when Adam and Eve fall from Eden. Here the message is to remain lowly wise, but in this epic Milton offers angels who use macrocosmic, air- and fire-fueled cannons that more resemble rockets as they move mountains. In the first outright epic written by a woman in England, *Order and Disorder* (1679), Puritan biographer Lucy Hutchinson presents a Miltonic, wounded mother earth challenged bodily again with the Noahic deluge. Hutchinson had already translated Lucretius' *De Rerum Nature* (likely the first English version of this dangerously materialist cosmological poem), but like Milton and unlike Cavendish, Hutchinson holds faithfully to a depiction of earth as subject entirely to God's Providence, even while she mixes into her account a material apocalypse and threads both of atomism and gynecology.

Of the three works, Margaret Cavendish's representation of earth would prove the most visionary from the perspective of modern science, as distinctions between the fields of meteorology and physiology grew, and geology emerged as a separate field (c. 1700); as the notion of an animate, mother earth at the center of the universe would eventually lose its appeal, even in literature; and as technology would drive innovation. The three texts above are evidence of the notable liberty afforded to the early modern imagination when expressing itself in decidedly literary form.

This is a liberty exercised playfully even by thinkers such as Hutchinson who were bent on working within Christian structures. Engaged with meteorophysiological speculation, this early modern liberty exceeds that exercised by thinkers centuries before and after. In early modern England, the rumbling of the meteorophysiological Titans of change could, as now, serve as a *memento mori*, signaling coming death, but it could also speak playfully to the notion of an animate earth, mother of all sublunary bodies, who is—along with each of her meteorophysiological children—a wonder of stability and eruption, vitality and mortality. Titanic rumbling could once, and might now, also speak of sublunary bodies interacting together, moved not by a human brain center but in response to change impacting variously at once a set of independent and interdependent actants across a system. Early moderns knew themselves to be complex, interactive, and elemental, appearing for a time in marvelous forms and motions, in fearful-pleasurable variety, and in constant exchange with all other beings and things, from animals to air. From this view of human being that finds its most ample representation in imaginative literature from the period, we might re-catch a new kind of enchantment enabling us to move into the future of clear and present meteorophysiological change from a place not at the center of the universe or with tightly held and precisely articulated accuracies rooted in apocalyptic fear of end times, but from our current intersection in this changing universe, which can be a place of representational abundance, choice, generosity and play and which can open us to wonders greater than our philosophies—any of them—have yet allowed us to dream.

Index

Note: Fictional characters are indicated by (fict.) following the uninverted name.

For Product Safety Concerns and Information please contact our EU representative GPSR@taylorandfrancis.com Taylor & Francis Verlag GmbH, Kaufingerstraße 24, 80331 München, Germany

Printed and bound by CPI Group (UK) Ltd, Croydon, CR0 4YY
14/09/2026
02216224-0010